XIX International Scientific and Practical Conference: Energy and Resource Saving XXI Century (ERS 2021)

IOP Conference Series: Materials Science and Engineering
Volume 1211

Orel, Russia
10 - 12 November 2021

ISBN: 978-1-7138-4670-3
ISSN: 1757-8981

Printed from e-media with permission by:

Curran Associates, Inc.
57 Morehouse Lane
Red Hook, NY 12571

CURRAN ASSOCIATES INC.
proceedings
.com

Some format issues inherent in the e-media version may also appear in this print version.

Institute of Physics
Dirac House, Temple Back
Bristol BS1 6BE UK

Phone: 44 1 17 929 7481
Fax: 44 1 17 920 0979

techtracking@iop.org

Additional copies of this publication are available from:

Curran Associates, Inc.
57 Morehouse Lane
Red Hook, NY 12571 USA
Phone: 845-758-0400
Fax: 845-758-2633
Email: curran@proceedings.com
Web: www.proceedings.com

TABLE OF CONTENTS

FOREWORD

Name of the conference: XIX International Scientific and Practical Conference "ENERGY AND RESOURCE SAVING - XXI CENTURY" (ERS 2021).

Location: Russia, 302026, Orel, st. Komsomolskaya d. 95, Department of Electrical Equipment and Energy Saving, FSBEI VO “OSU im. I.S. Turgenev ", tel .: (4862) 41-98-53, (4862) 41-98-30.

Date: November 10 - 12, 2021

Format: remote.

Founders:

- MINISTRY OF SCIENCE AND HIGHER EDUCATION OF THE RUSSIAN FEDERATION
- ORYOL REGIONAL GOVERNMENT
- ACADEMY OF ELECTROTECHNICAL SCIENCES OF THE RUSSIAN FEDERATION
- ADMINISTRATION OF THE CITY OF ORL
- PADERBORN UNIVERSITY (Germany)
- FEDERAL STATE BUDGETARY EDUCATIONAL INSTITUTION OF HIGHER EDUCATION "ORYOL STATE UNIVERCITY NAMED AFTER I.S. TURGENEV"
- FEDERAL STATE BUDGET EDUCATIONAL INSTITUTION OF HIGHER EDUCATION "I.I. POLZUNOV ALTAI STATE TECHNICAL UNIVERSITY"
- JSC "ORELOBLENERGO"
- LLC ORLOVSKIY ENERGOSBYT
- BRANCH OF PAO "MRSK CENTER" - "ORELENERGO"
- DEPARTMENT OF ELECTRIC EQUIPMENT AND ENERGY SAVING
- (Federal State Budgetary Educational Institution of Higher Education "I.S. Turgenev OGU")

Goals and objectives: development and popularization of the latest achievements of science, technology, as well as advanced experience in the implementation of energy and resource conservation, digital technologies in the electric power industry and electrical engineering; exchange of scientific and technical information for further expansion and strengthening of business contacts between domestic and foreign experts and firms.

Information about editors - heads of sections:

No. n/a	Section leaders	Section names
1	Turkov Andrei Viktorovich, Doctor of Technical Sciences, Professor of the Department of Building Structures and Materials ASI Federal State Budgetary Educational Institution of Higher Professional Education "Oryol State University named after I.S. Turgenev. I.S. Turgenev State University.	Problems of energy saving and safe operation of buildings, structures and urban areas
2	Kachanov Alexander Nikolaevich, Academician of AES RF, Honored Worker of Higher Education of RF, Doctor of Technical Sciences, Professor, Head of Department of Electrical Equipment and Energy Saving of FSBEI HE "Oryol State University named after I.S. Turgenev. D. in Economics, Professor of the Department of Electrical Equipment and Power Saving of the Orel State University named after I.S. Turgenev.	Energy efficiency of the electric power supply systems and the trends of their development

3	Kuvaldin Alexander Borisovich, Academician of Academy of Energy Sciences of RF, Honored Scientist of RF, Dr. Sci. "Professor, Dept. of Electrical Power Engineering of Industrial Enterprises and Electrical Technologies, MPEI, Moscow.	Energy saving electrical technological processes and installations
4	Barsukov Gennady Valerievich, Doctor of Technical Sciences, Professor, Department of Mechanical Engineering, TSU Oryol State University. I.S. Turgenev Oryol State University.	Energy saving engineering technologies and equipment
5	Nikolsky Oleg Konstantinovich, Honored Scientist of the Russian Federation, Doctor of Technical Sciences, Professor, Dep. "Professor of the Department of "Electrification of Production and Life", I.I. Polzunov Altai State Technical University, Barnaul.	Energy and resource saving in the agroindustrial complex
6	Podmasterev Konstantin Valentinovich, Honored Worker of Higher Professional Education of the Russian Federation, Doctor of Technical Sciences, Professor, Director of the Institute of Instrumentation, Automation and Information Technology, Turgenev State University of Orel. D. in Technical Sciences, Professor, Director of the Institute of Automation, Automation and Information Technologies of the Orel State University.	Information and metrological problems of the energy and resources saving
7	Polyakov Roman Nikolaevich, Doctor of Technical Sciences, Associate Professor, Head of Department of Mechatronics, Mechanics and Robotics, I.S. Turgenev Orel State University. I.S. Turgenev Orel State University.	Intellectual technologies and the automated control systems in the tasks of the energy efficiency increasing
8	Meshcheryakov Victor Nikolaevich, Honored Worker of Higher Education of Russian Federation, Dr. Sc. (Tech.), Professor, Head of Electric Drive Department, LSTU.	Digital technologies as a means for the energy efficiency improvement

Participants: teachers and staff of departments, graduate students, scientific and technical workers of problem laboratories of universities, research institutes, industrial enterprises, organizations and institutions working in the field of design, research and implementation of energy- and resource-saving, as well as digital technologies in various industries economy.

Contact faces:

Chernyshov Vadim Alekseyevich - Secretary of the Organizing Committee, Candidate of Technical Sciences, Associate Professor of the Department of Electrical Equipment and Energy Saving of Federal State Budgetary Educational Institution of Higher Professional Education "I.S. Turgenev OGU". E-mail: blackseam78@mail.ru
Korenkov Dmitriy Andreevich - Secretary of Program Committee, Candidate of Technical Science, Associate Professor of Electrical Equipment and Energy Saving Department, Federal State Budgetary Educational Institution of Higher Professional Education "I.S. Turgenev OGU". E-mail: di-mas.corenkov@yandex.ru
Lunev Roman Alexeevich - Secretary of the Technical Committee, Candidate of Technical Sciences, Associate Professor of the Department of Information Systems and Digital Technologies of Federal State Budgetary Educational Institution of Higher Professional Education "I.S. Turgenev OGU". E-mail: rolu@yandex.ru

IOP Conf. Series: Materials Science and Engineering 1211 (2022) 011002 doi:10.1088/1757-899X/1211/1/011002

Declaration on Peer Review
XIX International Scientific and Practical Conference "ENERGY AND RESOURCE-SAVING - XXI century" (ERS 2021)

The conference was held in the remote format on November 10-12, 2021 on the basis of FSBEI HEI "ORLOV STATE UNIVERSITY named after I.S. Turgenev".

• **Type of peer review: one-way blind, double-blind, open review, transparent review,/other**

All articles submitted for the conference underwent two levels of one-sided blind peer review.
The first level of one-way blind review was conducted by the section leaders in the following form:

A form of one-way blind peer review of the first level

XIX International Scientific and Practical Conference "ENERGY AND RESOURCE-SAVING - XXI century" (ERS 2021)

REVIEW OF THE ARTICLE/REPORT

Title of article / report:________________________________
Author / co-authors: ________________________________

1. The total value of the article/report:
a) novelty...☐
b) valuable contribution...☐
c) confirm existing views...☐
d) does not correspond to the subject...☐

2. Methods
a) are sufficient...☐
b) inaccurately described...☐
c) insufficient...☐

3. Experimental material and data
a) are sufficient...☐
b) insufficient...☐

4. Statistical evaluation
a) nenuzhnaya...☐
b) are sufficient...☐
c) is missing, but it is necessary...☐
d) based on false assumptions...☐

5. Illustrations
a) a sufficient quantity and quality...☐
b) insufficient...☐

6. Tables and figures
a) adequately and clearly represent the material...☐
b) must be improved...☐
c) inadequate or insufficient...☐

7. Interpretation of results
a) correct...☐
b) not correct...☐

8. Bibliography
a) correctly composed..☐
b) not in accordance with GOST..☐
c) incomplete; important links missing...☐
в) incorrect links...☐

9. Units of measurement
a) current (in the SI system)...☐
b) not suitable...☐

10. Abstract
a) satisfactorily...☐
b) does not characterize the essence of the work...☐
c) require a thorough revision..☐

11. The language of the article
a) good...☐
b) needs to be improved...☐
c) require a thorough revision..☐

12. The General opinion and suggestions regarding the publication of the article/report
a) very good, recommended..☐
b) well, it is recommended that, after adjustment...☐
c) acceptable after a thorough review...☐
d) bad - should not be published...☐

Comments, remarks:____
____

Reviewer:

The articles (reports) that successfully passed the first level of one-sided blind review were sent to the leading scientists working in the field of energy and resource conservation to pass the second level of one-sided blind review. When selecting a reviewer, the organizing committee took into account the correspondence of the scope of his scientific activity to the direction of the section in which the report (article) was submitted.

The second level of one-way blind review was conducted according to the form below:

Форма для одностороннего слепого рецензирования второго уровня
The XIX-th International Scientific Practical Conference "Energy and Resource Saving – XXI Century" (ERS 2021)

Title of the article:____
Author(s): ____

Mark with an "X" the items that best reflect your opinion about the article:
The content of the article: ____

The content of the article:

	Badly	Satisfied	Good	Fine
Compliance with the subject of the magazine*				
Relevance				
Originality				
Scientific and practical significance				

Quality of results and conclusions				
Construction and presentation of the material				

* if the answer is "bad", you can not answer the rest of the questions

The design of the article:

	Yes	No
There are no grammatical errors and typos in the article		
The equations and formulas are stated without errors		
The illustrations are high-quality and error-free		
The volume of the article meets the requirements		
The design of the article meets the requirements		

Conclusion on the possibility of publication:

Publication without corrections Publication with corrections	
Publication with corrections	
Publication is not possible	

Your comments and suggestions for making changes to the article:_____________

____

Reviewer:
(Full name, academic degree, academic title)
(Signature) Date:

- **Conference submission management system:**

On May 28, 2021 by order of the Rector of the I.S. Turgenev Moscow State University the website was created in order to provide operative management of the processes of organization and management of the conference (ERS 2021):

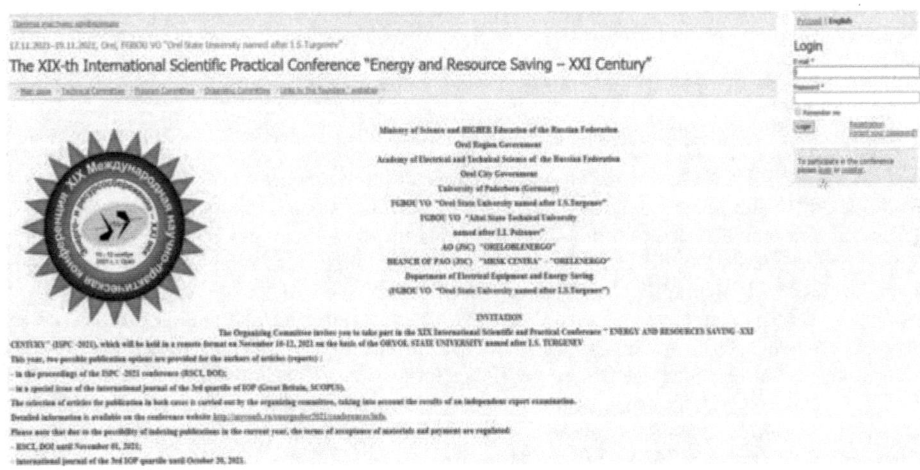

This site http://myconfs.ru/energosber2021/conferences/info provided information support to the participants of the conference, as well as the possibility of their communication with the organizers of the conference on all arising issues.

The site was also used for organizing the video conferences of the plenary and breakout sessions on November 10-12, 2021 in the ZOOM program.

- **Number of submissions received:**

The total number of papers (reports) received by the organizing committee was 66. 11 of them were formally rejected by the conference organizers.

- **Number of submissions sent for review:**

According to the results of the first level of reviewing, 25 out of 55 articles (papers) were sent for the second level of one-sided blind reviewing and the remaining 30 were recommended for the conference by the RINC (DOI).

- **Number of submissions accepted:**

The selected 25 papers (reports) were sent to the leading scientists in the field of energy and resource conservation to conduct the second level of unilateral blind review.
As a result, only 23 articles (papers) were recommended to be sent to IOP Publishers.

- **Acceptance Rate (Number of Submissions Accepted / Number of Submissions Received X 100):**

By a collegial decision of the participants of the plenary session held on 12.11.2021, out of 25 articles (papers) that underwent second level review, 23 were recommended for publication in the IOP journal, which is 92%;
- Average number of reviews per paper:
- 2 (two) reviews per paper for the IOP journal;
- 1 (one) review per paper for the RSCI Proceedings (DOI).
- Total number of reviewers involved:

The total number of reviewers involved for the first and second levels of one-way blind reviews is 22.

- **Any additional info on review process:**

At the preliminary stage of selection, the incoming materials were checked in the "Anti-Plagiarism" program according to the following criteria: "Originality", "Borrowing", "Self-citation" and "Citation".
All articles (reports) in the conference files have expert opinions on the possibility of their publication in the open press, drawn up collegially in accordance with the current legislation of the Russian Federation.

- **Contact person for queries:**

Russia, 302026, Orel, 95 Komsomolskaya str., Department of electrical equipment and energy saving of FSBEI VPO "OGU named after I.S. Turgenev", tel: (4862) 41-98-53, (4862) 41-98-30.
Chernyshov Vadim Alekseyevich - Secretary of the Organizing Committee, Candidate of Technical Sciences, Associate Professor of the Department of Electrical Equipment and Energy Saving of Federal State Budgetary Educational Institution of Higher Professional Education "I.S. Turgenev OGU". E-mail: blackseam78@mail.ru
Korenkov Dmitriy Andreevich - Secretary of Program Committee, Candidate of Technical Science, Associate Professor of Electrical Equipment and Energy Saving Department, Federal State Budgetary Educational Institution of Higher Professional Education "I.S. Turgenev OGU". E-mail: di-mas.corenkov@yandex.ru
Lunev Roman Alexeevich - Secretary of the Technical Committee, Candidate of Technical Sciences, Associate Professor of the Department of Information Systems and Digital Technologies of Federal State Budgetary Educational Institution of Higher Professional Education "I.S. Turgenev OGU". E-mail: rolu@yandex.ru

Methodology of fire risk analysis in electrical installations of a production facility

O K Nikolsky[1], Yu D Shlionskaya[1], M A Gabova[2], A N Kachanov[3], V A Chernyshov[3]

[1] Faculty of Energy, Polzunov Altai State Technical University, 46, Lenina ave., Barnaul, 656038, Russian Federation
[2] University college, Altai State University, 61, Lenina ave., Barnaul, 656049, Russian Federation
[3] Orel State University, 95, Komsomolskaya st. Orel region, Orel, 302026 Russian Federation

E-mail: gabova_ma@mail.ru, epb_401@mail.ru, kan@ostu.ru

Abstract . This article lists the principles of fire risk assessment and management of electrical installations, provides the main terms used in this subject area, and their meanings. The article also talks about factors affecting the magnitude of fire risk (risk-forming factors). Special attention is paid to the human factor. Formulas for probabilistic assessment of fire risks of electrical installations and an algorithm for assessing the fire hazard of electrical installations, as well as a scheme of the algorithm for calculating individual fire risk are given.

1. Introduction

Principles of fire risk assessment and management of electrical installations. Let us consider, within the framework of the current regulatory framework [1, 2], the basic concepts related to the term "fire risk" in relation to a certain set of electrical installations of a production facility:

- risk - a combination (product) of the probability of events and their consequences;

- permissible fire risk – its reasonable level, based on the socio-economic conditions of society at the present time;

- individual fire risk, which can lead to the death of a person as a result of exposure to fire hazards;

- social fire risk – the degree of danger leading to the death of a group of people as a result of exposure to fire hazards;

- a safe zone in which people are protected from exposure to fire hazards, or in which there are no dangerous factors;

- explosion - rapid chemical transformation of the medium, accompanied by the release of energy and the formation of compressed gases;

- explosion and fire hazard of the object of protection – the state of the object of protection, characterized by the possibility of fire and explosion;

- fire hazards (risk-forming factors), the impact of which can lead to loss of life, disability, poisoning and material damage;

- fire risk assessment – the process of comparing the real risk with its specified criteria in order to determine the significance;

- fire alarm system – a set of technical means designed to detect and notify of a fire, issue commands to turn on automatic fire extinguishing installations and turn on smoke protection installations;

- fire prevention system – a set of technical means that exclude the possibility of a fire at the protection facility (a protective shutdown device, which is a differential switch for leakage current, the value of which should not exceed 100 mA);

- fire risk management – a system of procedures and methods for managing emerging hazardous situations, identification, analysis, assessment, processing and monitoring of risk-related, to ensure loss reduction;

- fire risk management – an action taken to implement decisions within the framework of risk management;

- fire risk optimization - the procedure for finding the best risk value produced by:

a) at the design stage – minimizing its value with given material and financial resources;

b) at the operational stage - ensuring an acceptable (regulatory) risk value while minimizing (limiting) financial resources;

- a mathematical model of fire risk is the equivalent of an object that reflects in a formalized form its main dangerous properties in the form of patterns of cause-and-effect relationships.

2. Risk-forming factors of the fire situation and the assessment of the reliability of the "human factor".

The magnitude of fire risk (Rf) is influenced by a large number of various factors that characterize both the specifics of the specific operating conditions of a production facility (electrical installations of public and industrial buildings) and the specific features of danger, uncertainty, opportunities in which production activities are carried out [3].

The factors that contribute to the emergence and development of a fire situation at the facility will be called risk-forming factors (RFF). Considering that the number of man-made risk forming factors is quite large, we will limit ourselves to considering the so-called native risk-forming factors that are directly related to the subject area (diagnostics of fire safety of electrical installations).

In the literature sources [4, 5, 6, 7], the risk-forming factors of the human-machine system "Human-Electrical Installation-Environment" are considered, where the role of each component (electrical installation and environment) in the occurrence of dangerous technogenic situations was established. At the same time, it was noted that the technogenic safety of any object is determined by the reliability of its weakest (vulnerable) component, i.e. a person.

The negative impact of a person on the technogenic safety of an object will be denoted by the concept of "human factor", i.e. the totality of socio-economic abilities of a person, the degree of realization of which is due to the motivation and attitude of personnel to the process of work, their moral and material interest in high-performance work. The main manifestation of the negativity of the human factor is the following: a person in the course of his activity for one reason or another may make mistakes of various kinds. The property of a human operator to accurately perform his functions under given conditions of professional activity in time is interpreted as reliability, which is one of the components of professional fitness.

It is also necessary to take into account that in the process of life a person may find himself in an extreme situation when physical and psychological stress reach such levels at which the individual loses the ability to rational actions and decisions adequate to the current situation.

The human factor affects various activities of people, organizations, and economic objects. If the role of operator errors prevails in human-machine systems, then with an increase in the level of development of the social environment, the importance of erroneous organizational and managerial decisions affecting the stability of the functioning and sustainability of the development of relevant organizations increases.

The most common dangerous consequences are human errors – unintentional actions that go beyond the regulated boundaries, or failure to perform the necessary actions. Negative consequences of erroneous actions when working with technical objects are possible if the design does not provide for measures to neutralize them (Table 1) [8].

Table 1 - The main causes of erroneous actions of electrical personnel

Stages of production activity	List of reasons for ETP errors
Maintenance of electrical networks and electrical installations Actions and decisions of the operator Major repairs, reconstruction, manufacture of products and accessories Control of production works and events Personnel management, decision-making Socio-psychological aspects of the "human factor"	• Incorrect actions and unintentional mistakes in the production of works • Errors in operational decisions due to violations of regulations and instructions • Malfunction or absence of instruments (measurement and diagnostics) • Unclear formulation of the operational task • Violation of regulatory requirements • Inattentive inspection of equipment during its acceptance at the end of repair • Unsatisfactory involvement of management in ensuring work safety • Insufficient level of professional training • Indifferent attitude to the condition of the equipment • Deliberate deviation from operating and safety standards • Low motivation of professional behavior of the employee • Weak adaptive characteristics of an employee in extreme situations • Psychological and physical errors

The reason for the errors lies in the ratio of the capabilities of the personnel under the given conditions of activity and the loads (requirements) on it, which can be quantified using the "load-bearing capacity" model. A person's capabilities are related to his psychophysical characteristics, the degree of professional readiness to perform his functions. The conditions of activity in relation to the normative ones deteriorate with adverse environmental influences and irrational organization of work.

Figure 1 shows the classification of the causes of improper actions of electrical personnel leading to a fire in an electrical installation.

An erroneous decision is an incorrect unintentional execution or non-execution of a series of sequential actions due to an incorrect assessment of the flow of the regulated process. To assess the probability of errors, in particular, the method of personnel error analysis (Human Reliability Analysis - HRA) is used. Analyzing the decision-making process, it is possible to determine the causes of erroneous actions of personnel, which can become events that initiate an accident and the occurrence of a fire.

Quantitative assessment of the risk of electrical installation accidents is complicated by the uncertainties of the reliability characteristics of personnel, as already noted, the weak link of human-machine systems and their protection systems. In recent decades, methods for quantifying the reliability of the human factor have changed significantly compared to those traditionally used in calculating equipment reliability indicators.

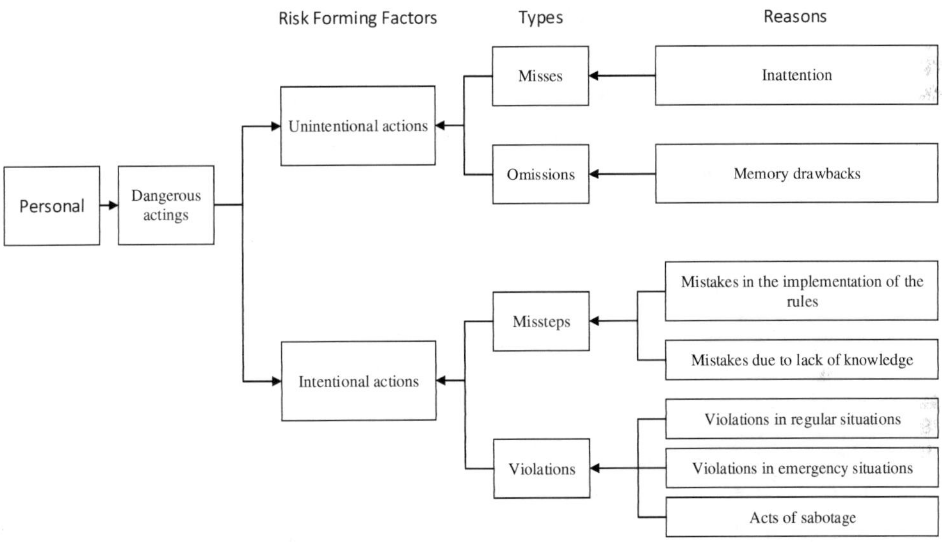

Figure 1 - Classification of the causes of improper actions of electrical personnel

The reliability of personnel in the maintenance of electrical installations of a production facility includes the following components [9]:

- psychosomatic, which is provided by training and depends on the properties of memory;
- cognitive, which is provided by the control and restoration of health;
- motivational, depending on the characteristics of the workforce, ensuring the formation of employee motivation.

An attempt to describe the properties of the operator using the mathematical formalism of the processes of hazard identification, development and occurrence of a fire situation is presented in Figure 2.

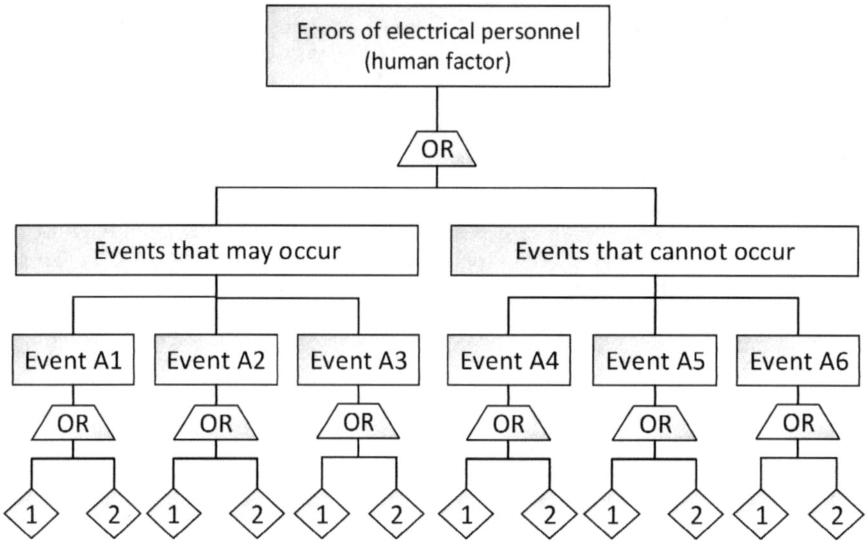

1 - dangerous reasons related to personnel; 2 - other dangerous reasons
Dangerous events caused by the "human factor": A1 - violation of the technological process;
A2 - improper use of equipment; A3 – the use of dangerous methods of work;
A4 - lack of control over the process of work;
A5 - violation of work and rest regimes, leading to errors of electrical personnel;
A6 - unsatisfactory maintenance of the workplace.

Figure 2 - Risk tree of erroneous actions of electrical personnel

3. Probabilistic assessment of fire risks of electrical installations.

Fire risk assessment is carried out by comparing the calculated risk values with its normative value established by Federal Law No. 123-FZ "Technical Regulations on Fire Safety requirements" [2].

The calculated values of fire risk are determined on the basis of:

a) analysis of fire safety of buildings; b) determination of the frequency of implementation of fire-hazardous situations; c) construction of fields of fire hazards for various scenarios of its development; d) assessment of the consequences of exposure of risk-forming factors of fire on people; e) availability of fire safety systems of buildings.

The determination of the calculated values consists in assessing the individual fire risk for people. The numerical expression of individual fire risk is the frequency of exposure to fire hazards (OFP) on a person in a building. The list of OFPS is established by Article 9 of the "Technical Regulations".

Fire hazard conditions of electrical installations at the facility in accordance with GOST 12.1.004-91 are determined by the formula:

$$Q_f = Q_R + Q_V + Q_P + Q_T \leq 10^{-6} \tag{1}$$

where Q_R–вthe probability of occurrence of a fire-hazardous mode (short circuit, overload, increase in transient resistance, etc.), 1/year;

Q_V – the probability that the value of the parameter (current, transient resistance, etc.) lies in the range of fire-hazardous values;

Q_P – the probability of failure of the protection device (electrical, thermal, etc.);

Q_T – the probability of the combustible material reaching a critical temperature or igniting it.

The data obtained on the estimated probabilities of fire occurrence are compared with the standard value of 1×10^{-6} per year, which is considered safe if the actual or estimated probability does not exceed the permissible one.

An objective indicator of the fire hazard assessment of electrical installations is the probability of a fire and its consequences (moral losses (loss of life), material and environmental damage. Moreover, this probability should not exceed the normative value.

The probability of fires from electrical installations of the object is determined by the formula:

$$P = \frac{n}{N}, \tag{2}$$

where n is the number of fires per year, N is the number of electrical installations in operation at the facility.

When analyzing the fire situation of an object, it is necessary to evaluate the reliability characteristics of the electrical installation.

Emergency fire-hazardous modes are determined by simulating malfunctions of elements in functional units of electrical equipment. Blocks are identified whose failure leads to the ignition of materials in the immediate vicinity.

The probability of ignition of an electrical installation is determined by the following expression:

$$Q = [1 - (1 - Q_R)(1 - Q_I]Q_f, \tag{3}$$

where Q_R – the probability of a fire, depending on the parameters of the reliability of the elements of the electrical installation of the EI;

Q_I – the probability of an ignition source arising due to the design features and manufacturing technology of the EI;

Q_f – the probability of failure of the EI protection device.

Value Q_R determined by the formula:

$$Q_R = 1 - \prod_{i=1}^{n}[1 - P^*_{i\text{К.З}}] \approx \sum_{i=1}^{n} P^*_{i\text{К.З}}, \tag{4}$$

where $P^*_{i\ \text{К.З}}$– the probability of ignition of the EC from a fire-hazardous component element of the i-th type; n is the number of types of elements. Then the probability of ignition of the product from transistors can be denoted as $P^*_{i\ \text{К.З}}$, from diodes $P^*_{2\ \text{К.З}}$, from capacitors - $P^*_{3\ \text{К.З}}$, ofrom transformers - $P^*_{4\ \text{К.З}}$etc.

Value$P^*_{i\ \text{К.З}}$ defined by the expression:

$$P^*_{i\text{К.З}} = 1 - \prod_{j=1}^{m}(1 - P_j) \approx \sum_{j=1}^{m} P_j, \tag{5}$$

where P_j – the probability of occurrence of an ignition source of a type i flammable element; m is the number of flammable elements of a certain type in an electrical product.

The value of P_j is determined according to the expression:

$$P_j = \lambda_j T P_{\text{КЗ/ОТК}} Q_{j\ \text{К.З}} Q_{j\ \text{К.М}}, \tag{6}$$

where λ_j – failure rate of the j-th element of the EI, 1/h;

T – the average duration of operation of the electrical installation;

$P_{\text{КЗ/ОТК}}$ – the probability of a short circuit in a fire-hazardous element in case of failure;

$Q_{j\ \text{К.З}}$ – the probability of ignition of the j-th element;

$Q_{j\ \text{К.М}}$ – the probability of ignition of structural materials located in the immediate vicinity of the EI.

The probability of occurrence of the ignition source of the electrical installation Q_E is determined by the formula:

$$Q_E = 1 - \prod_{k=1}^{L}[1 - P^*_{\text{К.Н}}] \approx \sum_{k=1}^{L} P^*_{k\ \text{К.М}}, \tag{7}$$

where $P^*_{\text{К.Н}}$ – probability of occurrence of the ignition source of the EI from the k-th type of production failures;

L – the number of types of failures. Then the probability of the ignition source of the EI from poor-quality solder joints can be denoted as $P^*_{1\ \text{К.М}}$, from short circuits of conductors – $P^*_{2\ \text{К.М}}$, from conductor breaks – $P^*_{3\ \text{К.М}}$, from violations of the contacts in the connectors – $P^*_{4\ \text{К.М}}$etc.

Value$P^*_{k\ \text{К.М}}$determined by the formula:

$$P^*_{k\ \text{К.М}} = 1 - \prod_{s=1}^{r}(1 - P_s) \approx \sum_{s=1}^{r} P_s, \tag{8}$$

whereP_s– the probability of occurrence of the ignition source of the EC from the s-th failure according to the k-th type of failure;

r –number of fire-hazardous failures by type k.

Probabilistic indicators of the occurrence of fire-hazardous failures of electrical installations are given in Table 1.

IOP Conf. Series: Materials Science and Engineering 1211 (2022) 012001 doi:10.1088/1757-899X/1211/1/012001

Table 1. Probabilistic indicators of the occurrence of fire-hazardous failures of electrical installations

Causes of failures	The probability of occurrence of an ignition source for various types of failures $P^{*}_{к.и}$
Poor-quality contact connections	4,0×10^{-2}
Conductor closures	0,19×10^{-2}
Conductor breakage	0,08×10^{-2}
Destruction of contacts	0,7×10^{-2}
Other refusals	0,03×10^{-2}
Total probability of failure	5,02×10^{-2}

Value $P^{*}_{к.и}$ determined by the formula (2).

The probability of failure of protection is calculated by the following formula:

$$P_{н.з} = k_1 k_2 , \quad (9)$$

where k_1 – the coefficient characterizing the protection of the EI from fire-hazardous modes;

k_2 – a coefficient that takes into account the presence or absence of a special fire extinguishing system in the EI. If there is such a system, the value k_2=0,05, in her absence k_2=1.

Value k_1 calculated by the formula:

$$k_1 = 1 - \frac{z}{N}, \quad (10)$$

where N - number of fire-hazardous modes (determined in the process of simulating malfunctions); Z - number of modes in which the EI protection is triggered (determined in the process of simulating malfunctions).

To assess the fire hazard of an electrical installation, it is necessary to develop a test program and methodology that takes into account conductive, insulating and structural elements and materials [10].

The algorithm for assessing the fire hazard of electrical installations is shown in Figure 3.

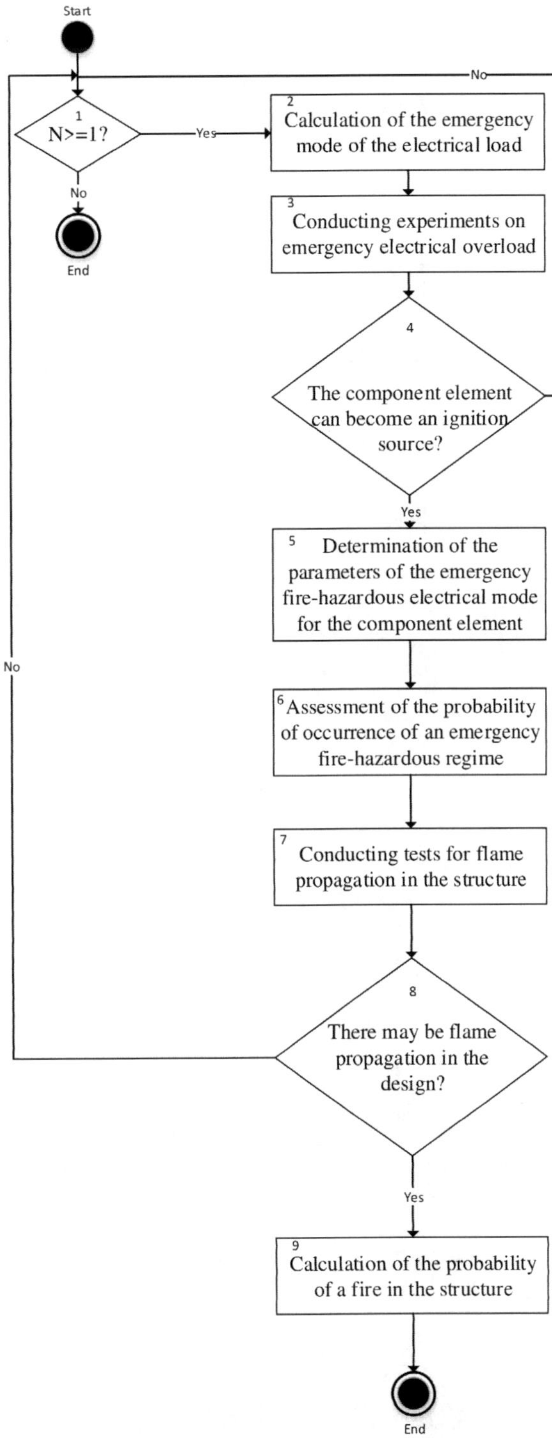

Figure 3 - Algorithm for determining the fire hazard of electrical installations of a production facility

In accordance with Article 93 of the "Technical Regulations", the value of an individual fire risk that can lead to the death of a person as a result of exposure to fire hazards for production facilities should not exceed 1×10^{-6}. For objects of the social sphere - the value of the individual fire risk is 1×10^{-7}.

Figure 4 shows a block diagram of the algorithm for calculating individual fire risk.

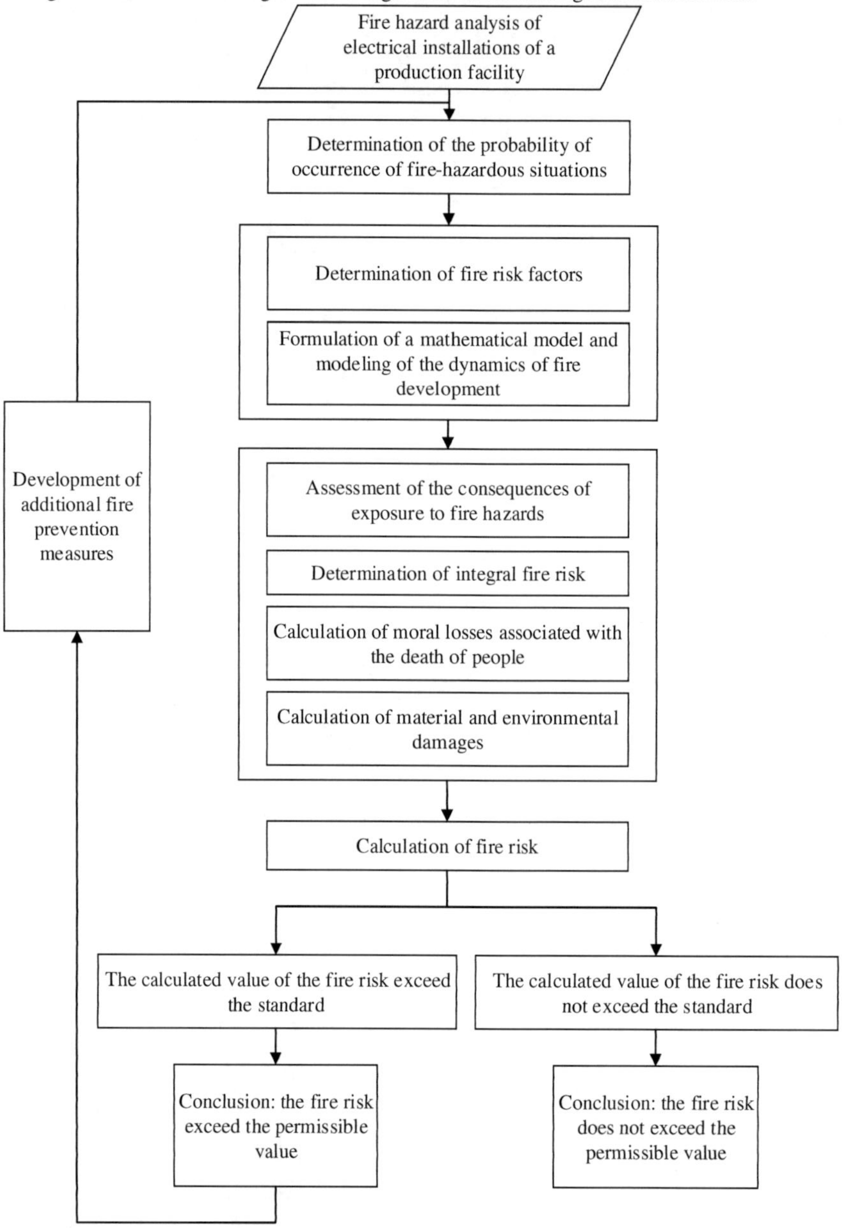

Figure 4 - Procedure for calculating individual fire risk

Conclusions

The state of fire safety of electrical installations of the infrastructure of cities and settlements in Russia poses a threat to national security, which led to the need to include the problem in the List of critical technologies approved by the President of the Russian Federation. The current situation is caused by a complex of problems of an economic, legal and technological nature, among which the most urgent are:

- lack of a unified concept of fire risk management of electrical installations of production facilities;
- lack of effective methods and tools for diagnosing the technical condition of the electrical facilities of the social sphere;

- imperfection of the regulatory legal provision of fire safety and lack of proper control by the supervisory authorities of the Ministry of Emergency Situations and gosgortehnadzor;
- the limited material and financial resources allocated to ensure technogenic safety, which led to significant depreciation (70%) of the fixed assets of electric power facilities.

References

[1] Federal Law No. 69-FZ of December 21, 1994 "On Fire Safety"
[2] Federal Law No. 123-FZ of July 22, 2008 "Technical Regulations on Fire Safety Requirements"
[3] Vorobyev, Yu.L., Malinovsky G.G., Makhutov M.A. Safety problems in emergency situations. - Risk theory and security technologies. - M., 2007. - 376 p.
[4] Germanenko, B.C. Conceptual foundations of safety management of electrical installations of buildings. - Polzunovsky almanac, - Barnaul: Alt. state Tech. un-t, 2004. - No.1-2. - pp. 113-120.
[5] Management of technogenic risks and optimization of the safety system of electrical installations of the agro-industrial complex infrastructure: abstract. dis. on the job. learned. step. Candidate of Technical Sciences: specialty 05. 20. 02 <Electrical technologies and electrical equipment in agriculture>. Shanygin I.A.-Ulan-Ude, 2020. -23 p.
[6] Nikolsky, O.K. Mathematical model of assessment and risk management of accidents in power supply systems / Nikolsky O.K., Kachesova L.Yu., Shanygin I.A. // Electrical equipment: Operation and repair - 2018. - No. 11.- pp. 72-77.
[7] Eremina, T.V. Fire hazard of electrical installations: models and forecasts / Eremina T.V., Shanygin I.A. // Scientific works of KubSTU ("SAFETY AND RESOURCE CONSERVATION IN THE TECHNOSPHERE" II International Scientific and Practical Conference, April 18-19, 2019). - Krasnodar, 2019. - No. 3. - pp. 572-582.
[8] Kulikova, E. Y. Fundamentals of life safety culture / E. Y. Kulikova // Scientific Bulletin of the Moscow State Mining University. - 2013. - No. 4.- pp. 51-55.
[9] Nikolsky, O.K. Modeling of technogenic risks of electrical installations of production facilities based on the analysis of human-machine systems / O.K. Nikolsky, Yu.D. Shlionskaya, I.A. Shanygin // Electrical Engineering. - 2018. - No. 12.- pp. 37-44.
[10] Fire safety in electrical installations / E. A. Andreev, A. S. Orlov, A. A. Levchuk, M. A. Marchenko // Life safety. – 2018. – № 1(205). – Pp. 52-56.

Methodology for comparing the energy efficiency of methods for regulating the performance of a blower station

Maksim Vecherkin[1,*], Anvar Sarvarov[1] and Irina Romanova[1]

[1] Magnitogorsk State Technical University named after G.I. Nosov, 455000, Magnitogorsk, Russian Federation

*Corresponding author e-mail: vecherkin@inbox.ru

Abstract. A methodology for comparing the energy efficiency of different methods of regulating the total productivity of a blower station is presented. The basic principles of electrical modeling of duct networks are shown for the mathematical description of their properties and characteristics. The results of a comparison of the energy efficiency of regulating methods for a single fan and a group of five fans operating on a common line are presented. The conclusion about the high energy efficiency of the combination of start-stop control with smooth control using axial guide vanes for a blowing station of five fans is obtained.

1. Introduction

Blower stations are a group of fans working in parallel on a common line to supply air to various process facilities.

In most cases the fans are fitted with high-voltage induction motors. These drives are initially designed as unregulated, making it impossible to vary the performance of the fans to the needs of the operation. This leads to overconsumption of electricity, which in some cases can be as high as 30% [1].

Reduction of unjustified power consumption is possible by adjusting the fan performance depending on the production needs. Consider a fan station with five fan units operating on a common main line and supplying air to three consumers of the same type (figure 1) [2].

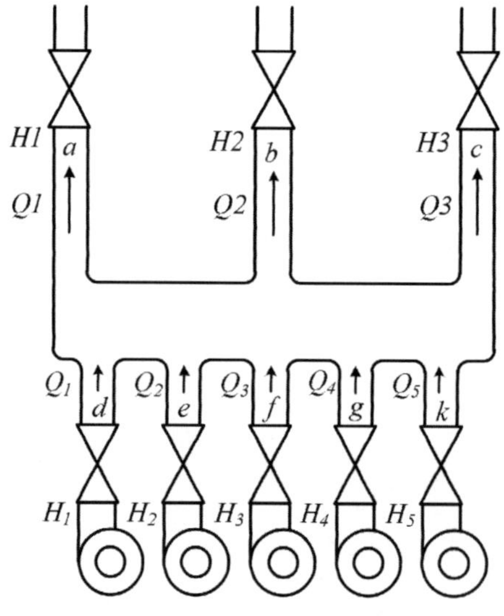

Figure 1. Structure of the blower station.

Each of the five fans generates the corresponding pressure H_1-H_5 and ensures the performance of Q_1-Q_5. From a common line, air is distributed to consumers, each of which has the ability to individually control the air flow rate by valves. In front of the valves of each consumer, pressures $H1$-$H3$ are created, the values of which should ensure the functioning of each consumer independently of the others. Thus, the blower station must provide performance in accordance with the required total flow rate, and also create pressure in front of consumers not less than the specified permissible values.

In general, the following methods of regulating the total capacity of the blower station are possible:

1) change of the network resistance by means of shut-off valves on the blower discharge side – throttling;
2) changing the speed of rotation of the fans – frequency control;
3) change in the number of operating fans – start-stop regulation;
4) changing the angle of rotation of the blades of the axial distributor of the fans – inlet-vane control;
5) a combination of two different methods – combined regulation.

The listed methods of regulation are not equivalent to each other both in terms of energy efficiency and in terms of material costs required for their implementation.

The most effective way is the frequency control of the fans. For its implementation, it is necessary to equip fan electric drives with frequency converters. However, the high cost of high-power high-voltage inverters prevents their widespread use. For fan stations, in most cases, the payback period of high-voltage inverters exceeds their service life [1].

Throttling is a standard means of regulating the air flow for each individual consumer. However, when fans are throttled, this method has low energy efficiency, which is confirmed by numerous theoretical and experimental studies [1, 3, 4].

Changing the angle of rotation of the axial distributor is essentially a throttling on the air intake side of the fan. The blades of the axial distributor then twist the air flow, which provides better control characteristics compared to throttling on the discharge side. For automatic regulation of the fan performance, it is necessary to equip the axial distributor with actuators that provide both the angular movement of the blades and the control of the angle of their rotation.

Start-stop regulation allows you to change the total performance in steps by changing the number of fans in operation. The implementation of this method requires the use of soft starters that neutralize the negative consequences of direct starting of powerful induction motors [2].

Combined regulation can be realized by a combination of different methods: frequency and inlet-vane control, start-stop and throttling, etc.

The problem is the choice of one method or another in each specific case, taking into account the characteristics of the technological object. In most cases, an experimental way of choosing a control method is impossible. It is rational to solve this problem theoretically, on the basis of research of a mathematical model describing the operation of fans and a network of air ducts.

2. Materials and methods

Two main approaches can be identified for comparing the energy efficiency of fans regulation: the graphical method and the electrical modelling method.

The graphical method has been developed in detail for single fans [1, 3, 5]. It makes it possible to determine any of the values required for a comparative analysis from the fan specifications and the known network parameters. Such values are: pressure H and productivity Q of each fan, mechanical power P on the motor shaft, efficiency η of fan. The advantage of the graphical method is clarity. A significant disadvantage of the method is the complexity of its implementation for a group of several fans [3].

The electrical simulation method is based on the analogy between the physical processes of electric current flow in branched circuits and air movement in duct networks. The different parts of the ventilation network are represented as equivalent electrical elements [6]. This makes it possible to represent any complex duct network as a virtual substitution network and to analyse it using Kirchhoff equations.

The work is based on the electrical modelling method. The basic principles of electrical modelling of ventilation networks are described in special literature [7]. The fan performance or air flow Q is equivalent to electric current, pressure H is equivalent to voltage, aerodynamic network resistance R is equivalent to electric resistance. The method is based on two obvious conditions: the balance of flow rates at nodes and the balance of pressure losses in closed loops. These conditions for a network without backpressure are written in general terms as follows:

IOP Conf. Series: Materials Science and Engineering 1211 (2022) 012002 doi:10.1088/1757-899X/1211/1/012002

$$\begin{cases} \sum_{i=1}^{n} Q_i = 0 \\ \sum_{j=1}^{m} H_j = \sum_{k=1}^{l} R_i \cdot Q_i^2 \end{cases} \tag{1}$$

where Q_i is the air flow through the network section; R_i is the aerodynamic resistance of the section; H_j is the pressure created by the fan in the closed circuit; n is the number of nodes in the circuit; m is the number of fans operating in the closed circuit; l is the number of sections with pressure losses in the circuit.

The difference between equations (1) and Kirchhoff's laws is that in the second equation the flow Q (current equivalent) enters in the second degree.

Based on the above, the duct network in figure 1 can be represented by the equivalent diagram in figure 2. The resistors in the diagram simulate the aerodynamic resistance of the network sections, the variable resistors are valves, the voltage sources are fans.

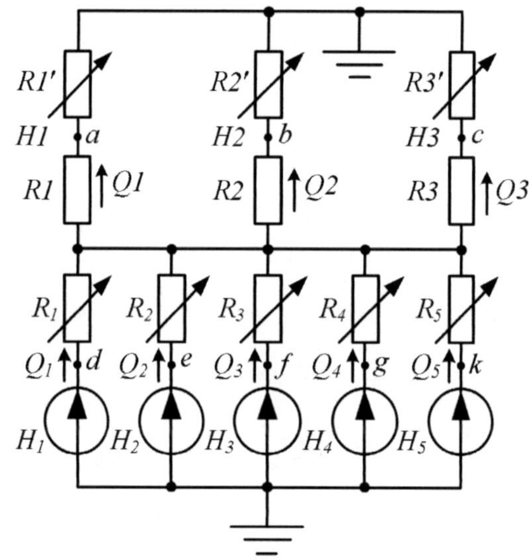

Figure 2. Electrical substitution diagram for a blower station.

For the equivalent circuit in fig. 2, according to conditions (1), a system of equations can be drawn up:

$$\begin{cases} Q_1 + Q_2 + Q_3 + Q_4 + Q_5 = Q1 + Q2 + Q3 \\ Q_1^2 R_1 - Q_2^2 R_2 = H_1 - H_2 \\ Q_2^2 R_2 - Q_3^2 R_3 = H_2 - H_3 \\ Q_3^2 R_3 - Q_4^2 R_4 = H_3 - H_4 \\ Q_4^2 R_4 - Q_5^2 R_5 = H_4 - H_5 \\ Q1^2 R1 + Q_1^2 R_1 = H_1 - H1 \\ Q2^2 R2 + Q_1^2 R_1 = H_1 - H2 \\ Q3^2 R3 + Q_4^2 R_4 = H_4 - H3 \\ Q1^2 R1 + Q_5^2 R_5 = H_5 - H1 \\ Q2^2 R2 - Q3^2 R3 = H3 - H2 \end{cases} \tag{2}$$

In real networks, the air pressures in the individual sections are often known and are determined from instrument readings. Usually the pressures upstream of the valves at points a, b and c (figure 1) are monitored, which is necessary for the functioning of metering and automatic control systems. Therefore, in equation system (2) the pressure drop at the valves is replaced by the corresponding pressures upstream of the valves $H1$, $H2$, $H3$.

The given system (2) cannot be solved in general terms. In order to implement numerical calculation methods it is necessary to set the air flow rates of consumers $Q1$, $Q2$, $Q3$, the aerodynamic resistances of all network sections $R1$, $R2$, $R3$, R_1-R_5, the pressures before the valves of each consumer $H1$, $H2$, $H3$. The unknowns are the capacities Q_1-Q_5 and the pressures H_1-H_5 of each fan.
The system of non-linear equations (2) has many solutions, most of which are unacceptable. These are, for example, solutions with negative values for the fan capacities or airflow rates of the consumers. In order to exclude these solutions, the condition of equal fan capacities must be set:

$$Q_1 = Q_2 = Q_3 = Q_4 = Q_5 \tag{3}$$

This condition provides a single solution to system (2), corresponding to the same positive performance values of all fans. In addition, condition (3) facilitates the search for the optimal, according to the criterion of minimum energy consumption, operating modes of the fans.
Fan performance and pressure are the main parameters that determine the fan's operating mode. The aerodynamic characteristics can be used to determine all the main energy parameters of the fan: effective power, mechanical power on the motor shaft, fan efficiency.
Aerodynamic characteristics are graphical dependencies of pressure on capacity, mechanical power on the fan shaft on efficiency and the fan efficiency on performance. These characteristics can be approximated with sufficient accuracy for engineering calculations by functions [3]:

$$H = a_1 Q^2 + b_1 \omega Q + c_1 \omega^2 \tag{4}$$

$$P = a_2 \omega Q^2 + b_2 \omega^2 Q + c_2 \omega^3 \tag{5}$$

$$\eta = \frac{QH}{P} = \frac{a_1 Q^3 + b_1 \omega Q^2 + c_1 \omega^2 Q}{a_2 \omega Q^2 + b_2 \omega^2 Q + c_2 \omega^3} \tag{6}$$

where H is total pressure of the fan; Q is performance; P is power at shaft; ω is motor speed; a_1, b_1, c_1, a_2, b_2, c_2 are coefficients depending on the blade angle of the axial distributor.
In order to obtain functions which enable analysis for all control methods, it is necessary to consider the dependence of the coefficients a_1, b_1, c_1, a_2, b_2, c_2 on the blade angle of rotation of the axial distributor. These dependencies may be represented in the form of analytical functions or by means of spline interpolation of data.
Thus, the system of equations (2) together with the condition (3) and approximating expressions (4)-(6) represent a mathematical model that allows analyzing various operation modes of the blower station in a wide range of performance changes at various control methods.

3. Results and discussions

Based on the presented mathematical model, a methodology for comparing the energy efficiency of control methods has been created. The essence of the methodology is as follows:

1. An equivalent electrical substitution diagram is created from the drawings of the duct network. The parameters of each circuit element are determined using the reference data.
2. A system of equations describing the electrical substitution diagram is compiled according to the rules (1).
3. The coefficients of polynomials (4)-(6) and their dependence on blades' angle of rotation are determined according to the known aerodynamic characteristics of fans: $a_1(\alpha)$, $b_1(\alpha)$, $c_1(\alpha)$, $a_2(\alpha)$, $b_2(\alpha)$, $c_2(\alpha)$.
4. Set the values of the air flow of each consumer $Q1$, $Q2$, $Q3$ and their corresponding pressures $H1$, $H2$, $H3$, based on the actual needs of the process facility.
5. The system of equations (2) is solved numerically. The unknowns in the system are the pressures $H1$-$H5$ and capacities $Q1$-$Q5$, which determine the operating points of the fans on the flow and pressure characteristics.
6. Calculate the regulator parameters corresponding to the operating points for the different capacity control methods:
   a) Aerodynamic damper drag (at nominal fan speed and open axial distributor);
   b) Fan speed (at open throttle and open axial distributor) as the positive root of equation (4)

$$\omega = \frac{-b_1 Q + \sqrt{(b_1 Q)^2 - 4c_1\left(a_1 Q^2 - H\right)}}{2c_1};$$

c) angle of rotation of the axial distributor blades (at rated fan speed and open throttle).
7. The power input (5) and efficiency (6) for each control method are calculated from the known operating points of the fans and the parameters of the controllers.
8. Steps 4-7 are repeated for different values of airflow by consumers.
9. Based on the results of the calculations, the following dependencies are plotted:
a) plots of the total power consumed by the blower station versus the total performance;
b) graphs of efficiency versus performance for each fan.
10. A comparative analysis is made of the power input and the efficiency for the operating conditions of the fans, the advantage of one or the other control method in terms of energy efficiency is concluded, and the limits for the acceptable fan performance ranges are evaluated. The limits of the ranges are determined by the value of the minimum allowable efficiency, which for the fans is equal to

$$\eta_{\min} = 0{,}9\eta_{\max} \tag{7}$$

where $\eta_{\max}$ is the maximum efficiency of the fan corresponding to the nominal performance.
Let's consider possibilities of this methodology with reference to single fan operation on a network without counter-pressure with unchanged parameters. The fan has high-voltage asynchronous electric drive with rated power 800 kW and aerodynamic characteristics typical for centrifugal fans [8]. Figure 3 shows the calculated dependencies of power consumption (figure 3, a) and efficiency factor (figure 3, b) in relative units at different methods of fan regulation (P – power, consumed by fan installation to create productivity Q, P_n – power, corresponding to the nominal performance Q_n). The graph for frequency control is shown with a typical frequency converter efficiency of 0.95.
The calculated dependencies in figure 3 show that for a single fan the frequency control has an advantage over other methods over a wide performance range. These results are in agreement with numerous theoretical and experimental studies, which confirms the validity of the methodology.

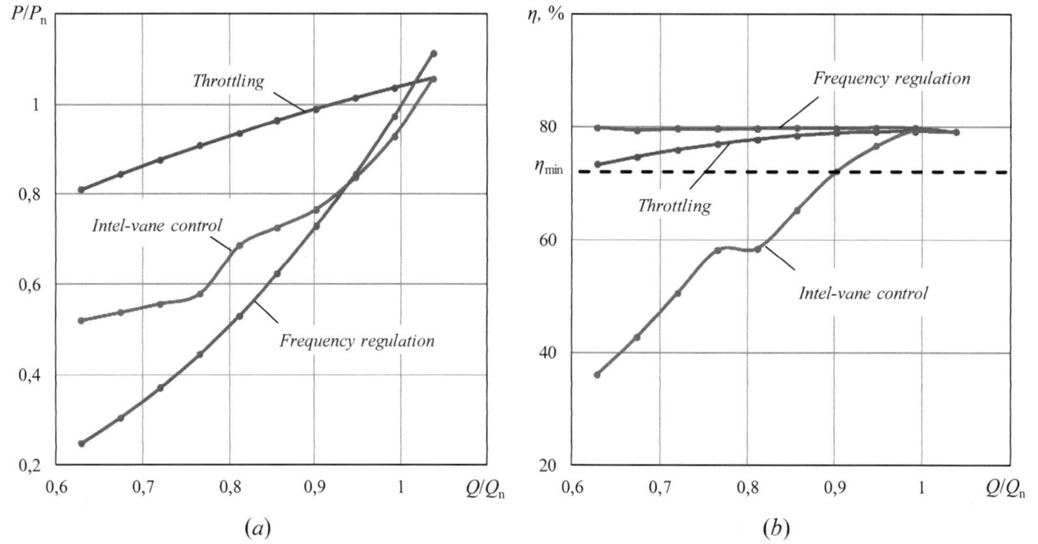

Figure 3. Calculated energy data for single fan operation: *a* – relative fan capacity from relative performance; *b* – Fan efficiency from relative performance.

Frequency control and intel-vane control are comparable in terms of energy efficiency with a small control range near the nominal capacity values. The advantage of frequency control is constant fan efficiency over the entire performance range (figure 3, a). Throttling has little effect on fan efficiency (figure 3, b), but is significantly inferior to other methods in terms of power consumption.
However, the results obtained for a single fan cannot be generalised to the case of a group of fans running in parallel on a common line. Figure 4 shows the graphs for frequency control of five fans and combined control by changing the number of running fans in combination with intel-vane control. The graphs are plotted for the case of simultaneous fan control, provided condition (3) is met.

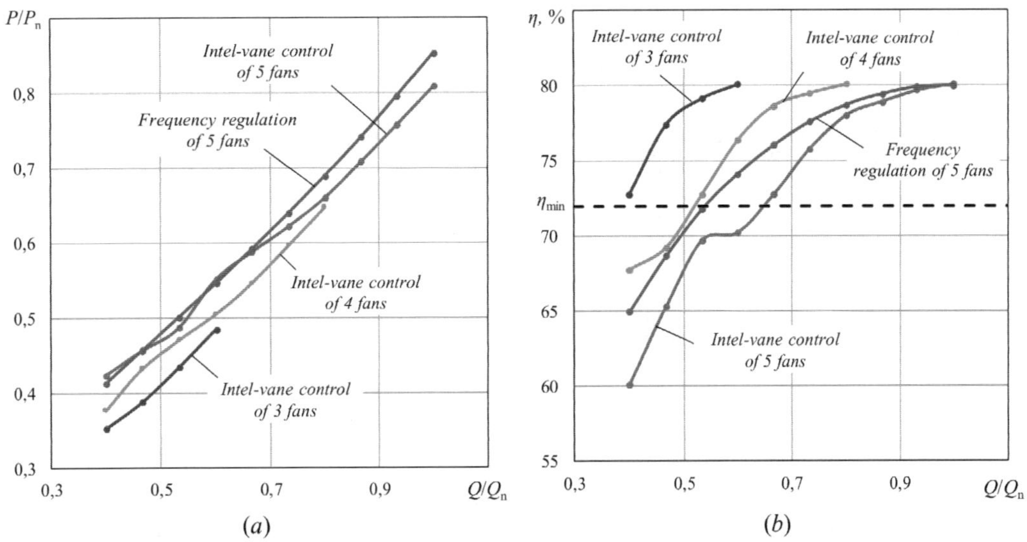

Figure 4. Calculated energy data for a group of fans: *a* – the relative total power of the blowing station from the relative total productivity; *b* – efficiency of one fan versus the relative performance of one fan.

The graphs show that when five fans are controlled simultaneously, frequency control has no advantage over intel-vane control in terms of power consumption. The advantage of intel-vane control increases when the combined control is used for small total performance. Turning off part of the fans both reduces the power input (figure 4, a) and increases the efficiency of the operating fans (figure 4, b).

4. Conclusions

The electrical modelling method is an effective tool for analysing ventilation networks. A mathematical model based on an equivalent electrical model and a mathematical description of the fans is a powerful tool for finding energy-efficient modes of operation for blower stations.
The given modeling technique allows solving a wide range of tasks related to fan operation in branched networks: analysis of energy efficiency of different performance modes; selection of fans for a network with given parameters; determination of pressure and flow values in separate sections; optimization of ventilation network operation [9], etc.
Examples of application of the methodology show that the established ideas about the advantage of frequency control of asynchronous electric drives are not always true for a group of fans running in parallel. In the case considered, frequency control loses out in terms of energy efficiency to the combined control method. This is not a general conclusion for all blower units. For each case, an analysis of the mathematical model is necessary, taking into account all relevant properties of the object in question [10].
The presented methodology can be adapted for other energy intensive technological objects: compressor, gas blower [11] and pumping stations taking into account the peculiarities of these objects.

5. References

[1] Leznov B.S. 2006 *Energy saving and variable electric drive in pumping and blowing installations*. (Moscow, Energoatomizdat), 360 p.
[2] Vecherkin M.V., Sarvarov A.S. 2006 *Study of methods to regulate the productivity of the fan station of Sheet Rolling Mill №10 of OJSC "Magnitogorsk Iron and Steel Works»*, Electromechanics, Vol. 62, №4, pp 63-67.
[3] Vahvahov G.G. 1975 *Operation of the fans in the network* (Moscow: Strojizdat), 101 p.
[4] Onishchenko G.B., YUn'kov M.G. 1972 *Turbomachinery electric drive* (Moscow: Energiya), 240 p.
[5] Vahvahov G.G. 1989 *Energy saving and reliability of fan systems* (Moscow: Strojizdat), 176 p.
[6] Glikman B.F. 1986 *Mathematical models of pneumohydraulic systems* (Moscow: Energiya), 368 p.

[7] Abramov F.A., Bajkov V.A., Frolov I.A. 1961 *Electrical modelling of mine ventilation networks* (Moscow, Gosgortekhizdat), 136 p.
[8] Solomahova T.S., CHebysheva K.V. 1980 *Centrifugal fans. Aerodynamic diagrams and characteristics: Handbook*, (Moscow: Mashinostroenie), 176 p.
[9] Himla J.W. 2009 *Optimized fan control in variableair volume HVAC systems using staticpressure resets: strategy selection and savings analysis*, (Texas A&M university) – Thesis for master of science degree. 70 p.
[10] Krupnikov A.V., Vanyashov A.D., Yanvaryov I.A. 2010 *Analysis of efficiency of air coolers on the basis of units with various number of fans* (Omsk: Omsk State Technical University) The Journal Omsk Scientific Bulletin №3 (93), pp 173-176.
[11] Kryachuk V.M., Vecherkin M.V., Shpon'ko A.A. 2011 *Diagnostics of Gas Blowers at OJSC "Magnitogorsk Iron and Steel Works" Coke Plant*, (New York: Allerton Press, Inc.), Coke and Chemistry, Vol. 54, №12, pp. 484-488.

Combustion of wood pellets with a high amount of fines

V K Lyubov, A N Popov and E I Popova

Northern (Arctic) Federal University named after M.V. Lomonosov,
Russia, 163002 Arkhangelsk, Severnaya Dvina emb., 17

E-mail: ev.popova@narfu.ru

Abstract. A comprehensive study of the efficiency of the 1.5 MW Arimax Bio Energy hot water boiler operating on wood pellets with a high amount of fines has been carried out. Fuel characteristics and its compliance with the quality requirements of Russian and European standards have been investigated. The components of the heat balance and emission of harmful substances have been determined. Thermovision study of the boiler was carried out. The comprehensive study showed that the boiler provides sufficiently high technical and economic performance and minimal emissions of nitrogen oxides and carbon monoxide. However, the high amount of fines in the pellets significantly increases the emissions of particulate matter and especially black carbon into the environment.

1. Introduction

The vital activity of mankind is accompanied by the formation of a large amount of waste, the volumes of which are growing rapidly and pose a serious threat to our time. Waste from the harvesting and processing of wood due to high moisture content, low energy density and extremely heterogeneous particle size distribution are classified as difficult-firing fuels [1-3]. A promising direction for upgrading these wastes is their granulation, which makes it possible to increase their specific calorific value by 2.5–3.5 times, improve their logistics by 3–4 times, and reduce by 6–10 times the cost of biomass transportation, while further improving the manufacturability of all operations. This direction has been intensively developing in Russia since the beginning of the XXI century. The energy potential of this direction in the field of bioenergy, as in the case of fossil hydrocarbons, surpasses those of any country in the world [4-6].

An effective direction for a comprehensive solution to energy and environmental problems while meeting heating needs is the use of modern automated heat-generating plants operating on biofuels. Therefore, a series of modern automated boiler houses operating on wood pellets was installed in the Arkhangelsk region.

Studies of the efficiency of combustion of wood pellets were carried out on an Arimax Bio Energy hot water boiler in one of these boiler houses.

2. Description of hot water boiler

The Arimax Bio Energy hot water boiler has two circuits, deaerated water circulates in the first circuit. The design mode for the power fluid of the primary circuit is 110/90 °C. The heating of the secondary circuit water (70/90 °C) takes place in a plate heat exchanger. The working pressure of hot water at the outlet of the boiler should not exceed 0.4 MPa, and the temperature should not exceed 115 °C.

The boiler screw feeder (Fig. 1) has a fuel re-ignition protection system, which is equipped with a fire detector and a sprinkler unit that is triggered when the screw surface temperature rises above 60 °C. The boiler feeder feeds the pellets to the profiled area of the furnace, where primary air passes through the gaps of the grate bars made of high-temperature cast iron with the addition of chromium, penetrating the fuel layer.

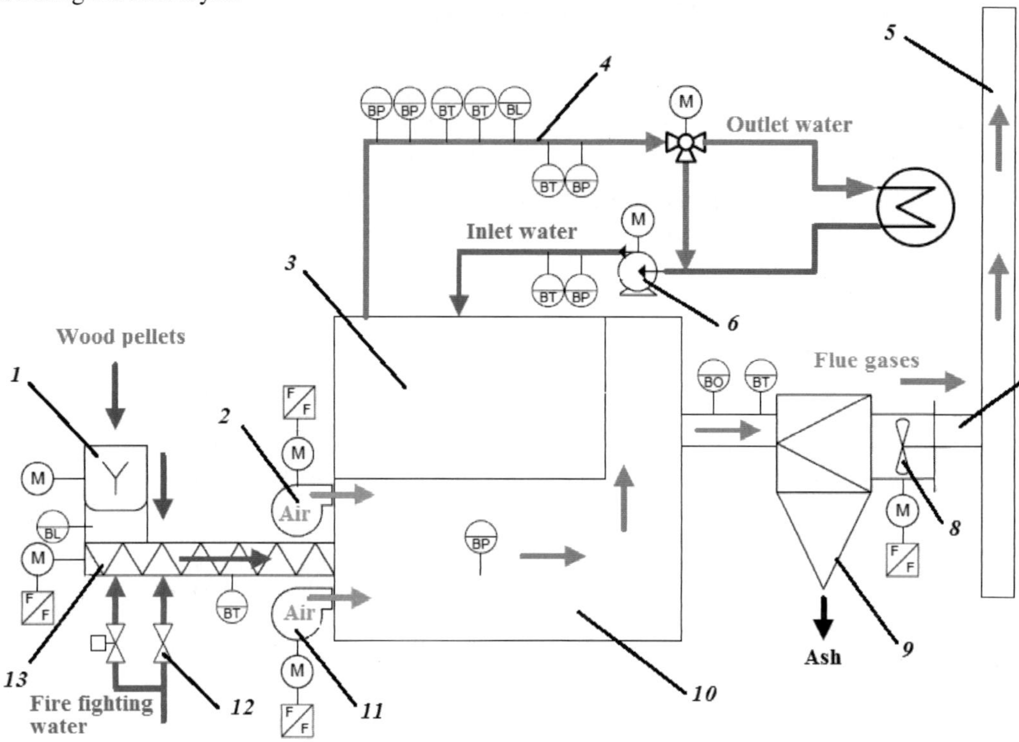

Figure 1. Arimax Bio Energy boiler layout: 1 – service fuel hopper; 2 – secondary air fan; 3 – boiler; 4 – hot water line to heat exchanger; 5 – chimney; 6 – circulating pump; 7 – gas duct; 8 – flue-gas fan; 9 – multicyclone; 10 – furnace; 11 – primary air fan; 12 – fire-fighting system; 13 – fuel feeder.

Secondary air is supplied to the above-layer zone of the furnace through nozzles with an inner diameter of 20 mm to ensure the afterburning of the combustible fuel components.

Combustion products provide heating of the boiler water of the primary circuit, making three pass in the fire tubes of the boiler, after which they are cleaned in a multicyclone (Fig. 1). The ash separated in the gas path is removed manually through the hatches located on the side walls of the boiler. The 1.5 MW Arimax Bio Energy multicyclone consists of nine elements. Particles of the caught ash are collected in a collecting hopper, which is tightly fixed to the outlet of the multicyclone. Combustion products by a flue-gas fan (with a 5.6 kW motor) are directed into a steel chimney with a diameter of Ø 325 × 8 mm and a height of h_{ch} = 10.5 m. Each boiler is equipped with an individual chimney. To ensure smooth regulation of the boilers performance, the motor of flue-gas fans, screws, fans of primary and secondary air has frequency control. The minimum heating capacity of hot water boilers Arimax Bio Energy is 20% of the nominal.

Arimax Bio Energy boilers are equipped with the necessary safety devices to protect them from boiling of the power fluid, and safety valves (Prescor) set to a pressure of 0.4 MPa. The automation control system, made on the basis of Siemens elements, provides: the required fuel-air ratio, the specified vacuum level in the furnace and the oxygen concentration in the combustion products, maintaining the temperature of the outlet water in accordance with the set temperature. The automation system is equipped with an Ethernet module and GSM modem, it provides the ability to remotely control the operation of the boiler room with a PC and a mobile phone, as well as controlling it through an Ethernet module using a PC.

The boiler is equipped with an individual control panel with automatic safety devices and automatic controls, performed on the basis of Siemens S7 200 microprocessors. Wall diagram is located on the outside of the control panel, showing the operation of the devices and is the center of the alarm.

Air intake fans of primary and secondary air to ensure the combustion process is carried out from the boiler room.

There is no possibility of controlling the resistance of the ash collector on the boiler, which does not allow an aerodynamic assessment of its degree of contamination. Monitoring the combustion process is carried out through the hatches located on the side walls of the boiler.

3. Materials and Methods

A series of tests with various modes of operation of the Arimax Bio Energy boiler was carried out during the energy study, when burning wood granules of the Sawmill 25.

Balance experiments were carried out in accordance with the requirements for production and operational testing of the second category of complexity [7].

Testo 350 XL gas analyzer was used in the study of the composition of combustion products. The pneumometric tube and Comark digital micromanometer were used to determine the flow of flue gases. The determination of velocity fields and flue gas flows was performed with a Pitot tube and micromanometer of Testo-435 precision instrument.

The external filtration method was used to determine the particulate matter concentration of the flue gases emitted into the atmosphere [7,8], while the filter holder of AFA was used to select the flow from gas ducts. The selection of dusting flow was carried out under conditions close to isokinetic. OP-442 TC impactor was used to measure and control the flow rate of selected combustion products. The calculation of the concentrations of the particulate matter in the flue gases and weight were carried out in accordance with the requirements [8].

The analysis of fuels was carried out with equipment of thermal analysis lab and IKA C2000 basic Version 2 calorimeter and LOIP-FT-216-25 cryothermostat.

The mechanical durability of wood pellets was determined in accordance with the standard EN 15210-1 using the NHP portable Tester 100 of the German company Holman (lignotester).

The study of the grain size distribution of fuel and focal residues was carried out using AS 200 Control and "029" analyzers [7].

The temperature of the outer surfaces of the main and auxiliary equipment of the boiler was determined using a pyrometer and the results of thermovision.

Processing of experimental data was carried out using a multi-module software and methodological complex [9].

4. Results

The main research results are given in Table. 1. The main research results are given in Table. 1. Wood pellets of Sawmill 25, having a fairly homogeneous particle size distribution, were fed to the combustion chamber of the boiler when carrying out balance experiments. The values of the main thermal characteristics of the pellets: moisture content as received basis W_t^r=6.62 %; ash content A^r=0.76 %; lower calorific value Q^r_i=17.34 MJ/kg. The mechanical durability of the pellets was 99.16%. It is necessary to note the increased ash content and the high amount of fines in the burning pellets. So the weights of the particles less than 3.15 mm was more than 4.5 %, which correspondively led to an increase in the bulk density of the fuel to 0.705 g/cm^3.

Thus, the burning pellets did not correspond to the class A1 on the ash content and the amount of fines, on the amount of fines they also did not correspond to the classes A2 and B. Pellets complied with the requirements of Russian and European standards for the remaining parameters.

Table 1. The main performance of the boiler.

Value	Symbol, dimension	Tests				
		No. 1	No. 2	No. 3	No. 4	No. 5
Heat capacity	Q, kW	651.3	976.9	1067.6	1239.8	1304.9
Outlet operating pressure of the water	$P_{o.p}$, MPa	0.21	0.21	0.22	0.23	0.23
Inlet water temperature	$t_{i.w}$, °C	96.7	95.1	92.8	92	91
Outlet water temperature	$t_{o.w}$, °C	104.5	106.7	108.8	108.9	109.0
Air flow rate primary/secondary	V_{pr}/V_{sec}, Nm^3/h	392/682	568/988	504/877	577/1005	599/1042
Furnace draft / before ash collector	$S_f/S'_{a.c}$, Pa	47/135	60/145	53/200	58/210	54/250
Draft loss	S_b, Pa	130	85	147	152	196
Flue gas temperature	$\vartheta_{f.g}$, °C	148.1	175.5	194	196.5	206.9
Excess air in flue gas	$\alpha_{f.g}$	1.58	1.54	1.27	1.26	1.24
Heat loss: flue gas	q_2, %	7.41	8.94	8.61	8.68	9.12
incomplete combustion	q_3, %	0.07	0.03	0.02	0.01	0.01
carbon	q_4, %	0.95	0.95	0.95	0.95	0.95
external	q_5, %	5.99	3.99	3.65	3.15	2.99
Gross efficiency of the boiler	η_{gross}, %	85.56	86.06	86.74	87.19	86.91
Total fuel consumption	B, kg/h	157	235	254	294	310
Emission of NO_x	NO_x, mg/MJ	40	46	49	46	45
Emission of CO	CO, mg/MJ	71	35	31	15	14
Particulate matter emission	PM, mg/MJ	-	-	-	60.1	-

Analysis of the state of the gas path of the hot water boiler showed that the boiler has a high aerodynamic density of the lining. Inleakage of cold air in the section of the gas duct ash collector – flue-gas fan does not exceed the standard value ($\Delta\alpha_{a.c+flue} < 0.05$). However, the aerodynamic density of the gas paths should be periodically monitored, especially the zone of connection of collecting tanks to multicyclones.

The results of the experimental studies carried out have shown that the Arimax Bio Energy boiler is capable of providing the rated load with normal thermal characteristics of wood pellets.

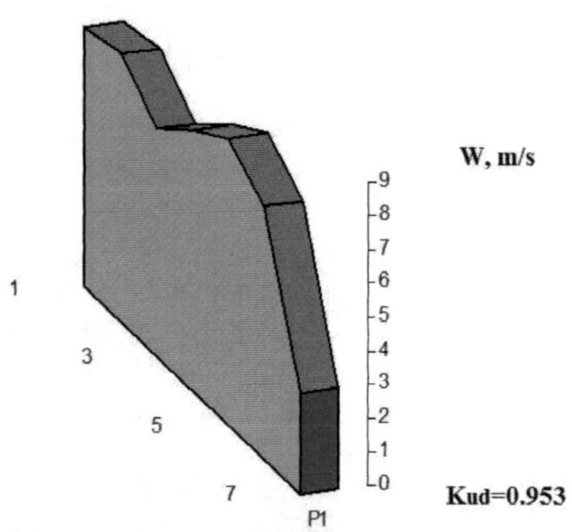

Figure 2. Changing the air velocity of the secondary air in the fan inlet.

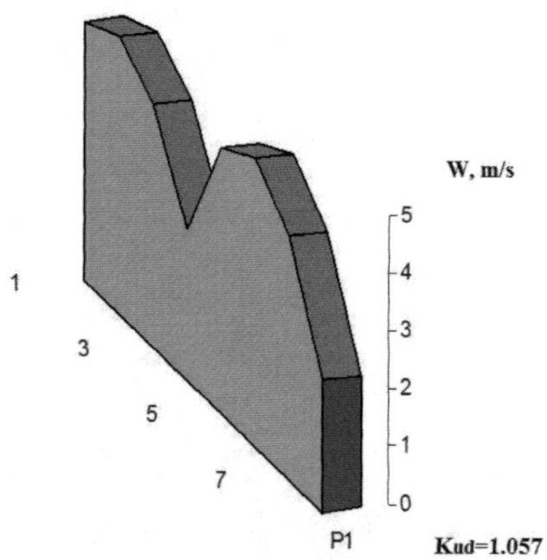

Figure 3. Changing the air velocity of the primary air in the fan inlet.

Balance experiments were carried out in the range of boiler loads from 60 to 100% of the nominal, while the water temperature at the outlet from the boiler varied in the range of $t_в$ = 104.5–109.0 °C.

During the energy study, the excess air ratio of the supplied air was 1.18–1.52; and the fraction of primary air (0.45–0.55) was less than the fraction of secondary air, which is quite reasonable for wood pellets.

The ratio of primary to secondary air was 0.37 / 0.64, respectively. The boiler provides a high combustion efficiency of carbon monoxide even at low values of the excess air ratio at the outlet from the furnace, and its operation at high values of the excess air ratio ($\alpha_f'' > 1.52$) is accompanied by an increase in carbon monoxide emissions ($K_{CO} \geq 176$ mg/Nm^3). Changes in the velocities of primary and secondary air in the fan inlet are shown in Figures 2, 3.

5. Discussion

Heat losses due to incomlete combustion of the fuel were q_3 = 0.01–0.07 %. The boiler gross efficiency varied in the range 85.56–87.19 %. The specific consumption of fuel for the production of 1 Gcal was 163.7–166.3 kg of reference fuel/Gcal.

The water temperature at the boiler inlet was 91.0–96.7 °C during the energy tests. The gas-water heat exchanger provided insufficiently deep cooling of the combustion products, which caused a decrease in the efficiency of the boiler.

The decrease in the boiler efficiency also caused a significant amount of fines in the fired fuel. For comparison, at another boiler house in the Arkhangelsk region, similar tests of Arimax Bio Energy boilers with a nominal capacity of 1.5 MW were carried out [10]. The boiler efficiency in this case reached 90.38%, since wood pellets with a low amount of fines were burned in the boiler furnace.

During the tests of the boiler house, a significant discrepancy between the values of the indicating devices for vacuum in the furnace of the Arimax Bio Energy boiler and the values displayed on the control panels was revealed.

The draft loss of the boiler along the gas path depends on its load, operating modes and characteristics of the fuel. When the heat output of the boiler was changed in the range of 60–100% of the nominal, its draft loss was 70–250 Pa. The draft loss of the multicyclone could not be determined due to the absence of fittings on the gas ducts after it.

The boiler is cleaned at least once a year, and the multicyclone - once a month. Critical parameter for cleaning heating surfaces is an increase in the flue gas temperature to values of 250 ° C and above (at a nominal heat capacity).

To maintain the degree of gas purification in the multicyclone at a high level, it is necessary to control the density of connection of collecting hoppers to the ash collector, which will avoid secondary entrainment of fly ash and an increase in energy consumption for draft. The noted regularities take place during the combustion of pellets with a uniform grain size distribution and a small amount of particles. Violation of the normal technological process when pelleting wood raw materials, especially an increase in its moisture content above the permissible level, leads to a low quality of the pellets and their subsequent self-destruction. The supply of such pellets or pellets with a high amount of fines to the furnaces of Arimax Bio Energy boilers can significantly reduce the efficiency and reliability of their operation due to a significant increase of carbon loss. Based on this, the incoming quality control of the pellets should be organized.

Analysis of the thermal operation conditions of the Arimax Bio Energy boiler showed that the heat loss with flue gases was 7.41–9.12%. At the same time, an increase in this loss is noted with an increase in heat load.

Carbon loss was q_4 = 0.95%. The combusted fuel contained high amount of fines, which caused the removal of particles that did not have time to burn out in the furnace volume of the boiler, while the content of combustible components in the fly ash was $C^c_{f.a}$ = 21.11 %. The content of combustible substances in the slag has increased significantly (C^c_{slag} = 43.48 %).

Most of the boiler piping has high-quality thermal insulation. However, there are pipelines where there is no thermal insulation (Fig. 4), which contradicts the requirements of regulatory documents [11-13].

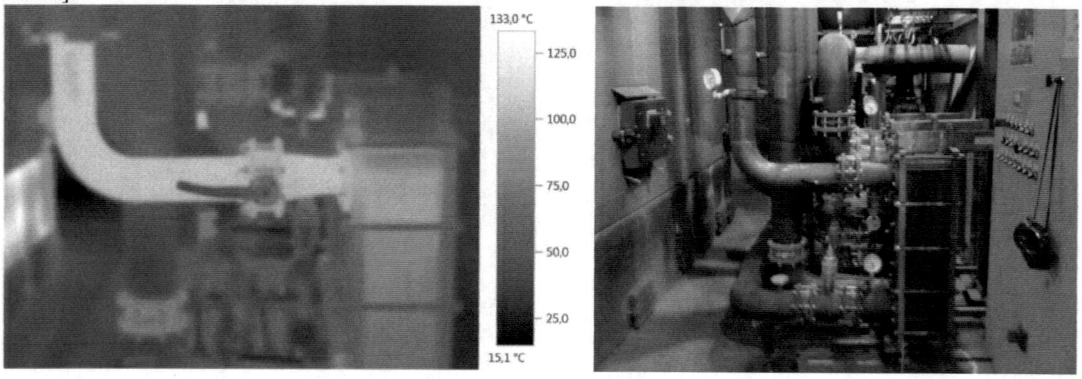

Figure 4. Boiler room thermograms.

The reduced overall parameters of the Arimax Bio Energy boiler and the satisfactory quality of the lining and heat-insulating materials made it possible to reduce external heat losses, which cannot be determined using the traditionally used method [14]. Based on this, the value of this loss was determined on the basis of the results of measurements of the temperature of the outer surface of the boiler casing using a pyrometer and thermovision taking into account the overall characteristics and the ambient temperature. The results of thermovision made it possible to identify areas of the enclosing structures where the temperature does not meet the requirements of the current standards (Fig. 5-8). The calculated values of external heat losses were q_5 = 2.99–5.99 %, large values of these losses correspond to the boiler operation modes with reduced loads.

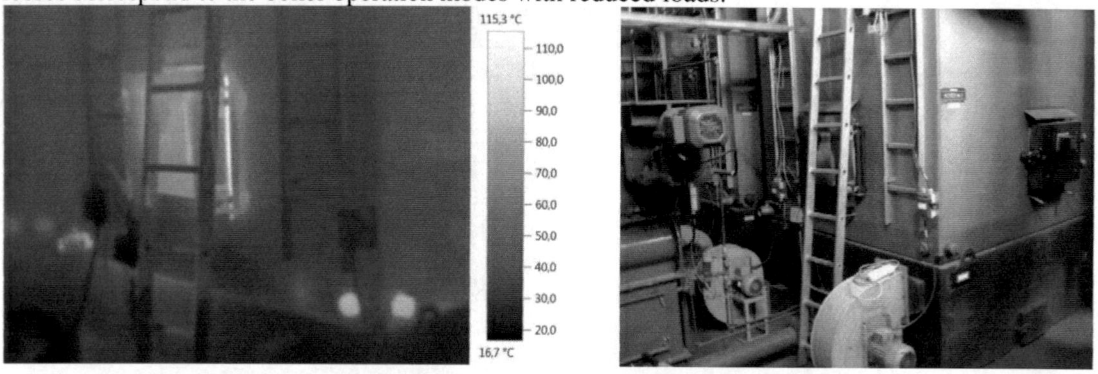

Figure 5. Thermograms of the front wall of the boiler.

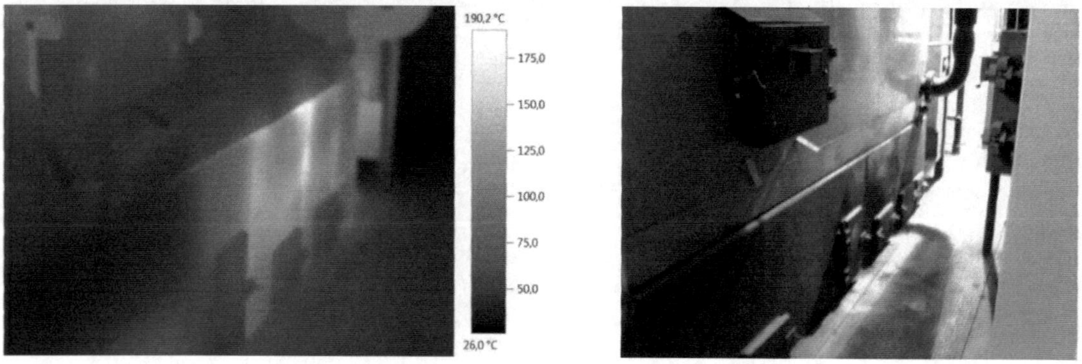

Figure 6. Thermograms of the right side of the boiler.

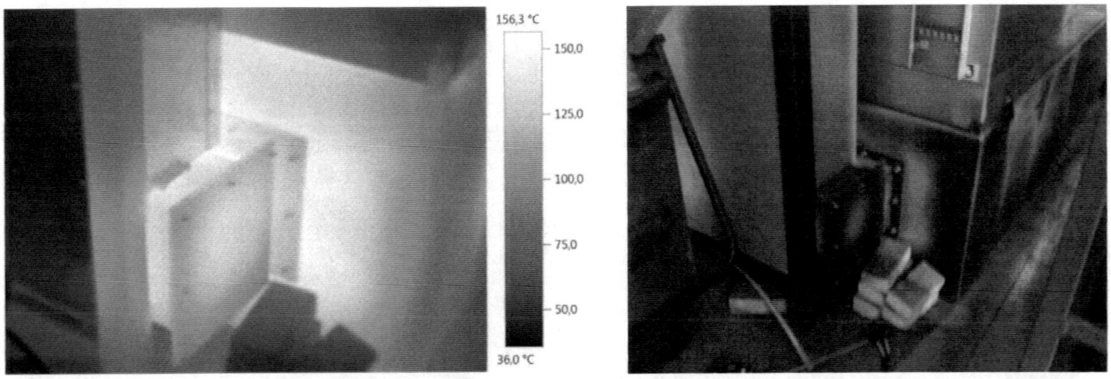

Figure 7. Thermograms of the rear wall of the boiler.

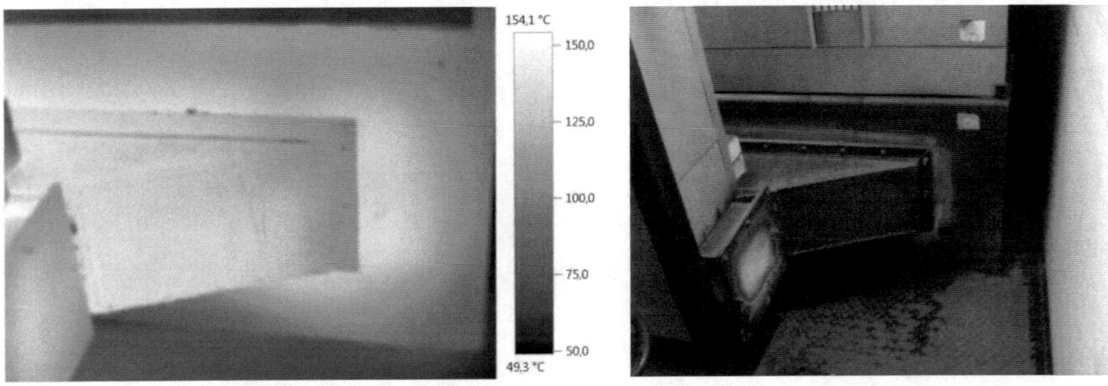

Figure 8. Thermograms of the gas dust after the boiler.

The total fuel consumption for the specified range of heat loads was B=157–310 kg/h. It should be noted that the panel meters for fuel consumption in the boiler room give underestimated values. The data on the consumption of wood pellets obtained from the boiler panel meters cannot be used for calculations. Objective data on fuel consumption can be obtained only on the basis of the inverse heat balance equation, taking into account the recommendations [14,15].

Low values of nitrogen oxide emissions from 40 to 49 mg/MJ are explained by a moderate level of maximum temperatures, as well as a two-stage fuel combustion scheme.

Sulfur dioxide was absent in the flue gases.

The range of changes in carbon monoxide emissions during the balance tests was 14–71 mg/MJ. The results of the experiments have shown that the Arimax Bio Energy boiler provides efficient combustion of combustible fuel components at an oxygen concentration in the combustion products at the output of the boiler of 4.0–6.0 % (at standard boiler density).

Studies of the particulate matter have shown that the emissions of particles have increased values (60.1 mg/MJ), which is associated with a high amount of fines in the fuel. Accordingly, the measurements performed also showed increased values of soot particle emissions (12.69 mg/MJ), which is explained not only by the high amount of fines in the burned pellets, but also by the high content of combustible substances in the fly ash.

Studies of the grain size distribution of fly ash caught in the ash collector showed that it has a high degree of polydispersity (n=0.740; b=0.0478) and belongs to finely dispersed materials (Fig. 9). The fly ash is dominated by fractions with a particle size of less than 250 microns, which account for more than 98%. The results indirectly suggest a high separation capacity of the ash collector.

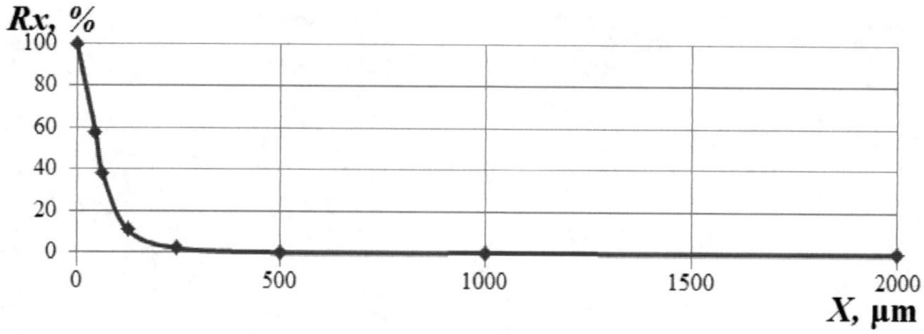

Figure 9. Fly ash grain size distribution.

Studies of the distribution of combustible substances by fractions in fly ash have shown that particles with a size of 125 microns and more have the maximum combustible content (Fig. 10). However, their mass fraction in fly ash is small. Therefore, the decisive influence on carbon loss with slag and fly ash is exerted by the content of combustible substances in particles less than 125 microns in size (Fig. 11).

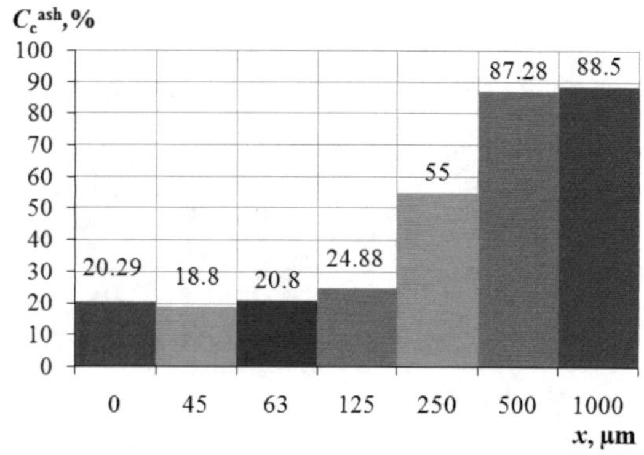

Figure 10. Fractional content of combustible substances in fly ash.

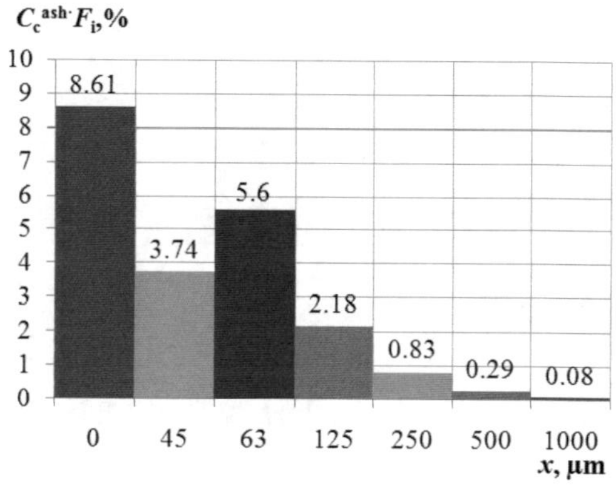

Figure 11. The content of combustible substances in fly ash, taking into account the mass fractions of various fractions.

6. Conclusion

The results of a comprehensive energy study showed that the Arimax Bio Energy boiler, operating on wood pellets, despite the previously mentioned drawbacks, has sufficiently high technical, economic and environmental performance, but there is significant potential for their further improvement.

References

[1] Golovkov S I, Koperin I F and Naydenov V I 1987 *Energy use of wood waste* (Moscow: Lesnaya promyshlennost',) p 224
[2] Zhuchkov P A, Goflin A P and Saunin V I 1991 *Heat engineering for pulp and paper industry* (Moscow: Ekologiya) p 352
[3] Koperin I F and Golovkov S I 1985 *Boiler plants of timber industry* (Moscow: Lesnaya promyshlennost') p 207
[4] Gerasimov Y and Karjalainen T 2011 *Biomass Bioenergy* **35** 1655
[5] Berdin V K 2020 *Eur. J. Political Econ.* **15** 106
[6] Porfiriev B N and Roginko S A 2016 *Her. Russ. Acad. Sci.* **1** 433

[7] Trembovlya V I, Finger E D and Avdeeva A A 1991 *Thermotechnical tests of boilers* (Moscow: Energoatomizdat) p 416
[8] 1987 *Collection of Methods of determination of harmful substances concentrations in industrial emissions* (Leningrad: Gidrometeoizdat) 248
[9] Lyubov V K and Dyachkov V A 1998 *Proc. of 2nd Russian national Conf. on heat exchange* (Moscow: MEI) pp 225–228
[10] Lyubov V K, Mueller O D and Popov A N 2012 *XIV Minsk international heat and mass transfer forum* **303**
[11] 1993 *Rules for the design and safe operation of steam boilers with a steam pressure of not more than 0.07 MPa (0.7 kgf/cm^2), hot water boilers and water heaters with a water heating temperature not higher than 388 K (115 °C)* (Moscow: NPO OBT) p 185
[12] 2001 *GOST 30735-2001. Hot water boilers with heat capacity from 0.1 to 4.0 MW. General technical conditions* (Moscow: Izdatel'stv standartov) p 18
[13] 1992 *Rules for the technical operation of heating boiler houses* (Moscow: NPO OBT) p 87
[14] 1998 *Thermal design of boilers (Standard method)* (Saint-Petersburg: NPO CKTI) p 256
[15] Lyubov V K and Lyubova S V 2017 *Increasing of biofuels usage efficiency* (Arkhangelsk: NArFU) p 533

IOP Conf. Series: Materials Science and Engineering 1211 (2022) 012004 doi:10.1088/1757-899X/1211/1/012004

SRM with 8/6 magnetic system topology for electric drive of mine battery electric locomotives: design and modelling

V N Antipov, A D Grozov and A V Ivanova

Institute of Silicate Chemistry RAS, 2, Makarova emb., Saint Petersburg, 199034, Russia

E-mail: bht@mail.ru

Abstract. The paper deals with the development and simulation results of the switched reluctance motor for electric drive of mine battery electric locomotives instead of the DRT-14 DC motor. The switched reluctance motor parameters are obtained based on numerical calculation of the magnetic field by QuickField program and are embedded in MATLAB/Simulink model of a switched reluctance motor created for 8/6 magnetic system configuration. SRM-14-615 has the mechanical characteristics as DC motor DRT-14 and meets the required operating modes as part of the AM8D mine battery electric locomotive. Also two types of models using methods and techniques of the artificial intelligence theory are presented: a model with a fuzzy control system in the Fuzzy Logic Toolbox package and a model with a neural network control system in the Neural Network Toolbox package.

1. Introduction

Switched Reluctance Motor (SRM) drives have been used for many years in applications, where design simplicity was primary important. The theory of a rotating magnetic field and well-developed methods for calculation of rotating electric machines are not applicable to SRMs. The literature notes the complexity of solving the problem of SRM with high technical indicators creating. Numerous studies have almost overcome this drawback [1]-[3].

We believe the SRM development has to be based on the principles of block-modular motor design. It means that a basic industry-developed motor prototype is selected and all necessary components and parts (housing, shaft, bearings, shields, etc.) are saved, with the exception of the developed stator coils, stator and rotor packages, as well as a small shaft revision. Thus, in block-modular motor design, the constraints are the outer diameter of the stator package and the inner diameter of the rotor package, the dimensions of which must be saved when designing the stator and rotor iron sheets. The principles of block-modular motor design ensure a minimum of production and financial costs [4].

The SRM design procedure differs significantly from the traditional one and includes three interrelated design stages [4], [5]. The authors have developed their own algorithms and calculation programs based on EXCEL, QuickField, and MATLAB technologies for each of the design stages.

The inclusion in the design procedure of typical mathematical blocks of visually-oriented modeling MATLAB matrix system with the Simulink and SimPowerSystems extension packages opens up additional opportunities for effective design [5]-[6].

This paper presents the results of a study in the Matlab/Simulink environment of a four-phase Switched Reluctance motor with the topology of the magnetic system 8/6 for a mine battery electric locomotive. For study of this motor topology three Simulation block diagrams have been created for controlling the speed of rotation, battery including those created with methods and techniques of the theory of artificial intelligence. The created models were verified in various modes of a wide range of motor speed control.

2. Materials and methods

The electric motor SRM-14-615 (14 kW, 145 V, 615 min^{-1}) is designed for electric drive of mine battery electric locomotives AM8D and 2×AM8D with a coupling weight of 8 tons and meets the requirements for the currently used electric motor DRT-14 (14 kW, 130 V, 615/1845 rpm). The prototype is designed and founded on the principles of block-modular motor design based on DC motor 4PF200L. The *n*-sided frame of electric motor 4PF200 is used as a stator SRM-14-615 and the axial length of the rotor package is limited to 200 mm. The transition to an n-face frame allows you to effectively implement a variant of a four-phase electric motor m=4, N_s=8, N_r=6. The width of the stator (rotor) tooth in the angular version is considered optimal – 20.3 (23.7) degrees [5]. To select the rotor diameter, calculations were performed for four of its values in the range of 210–261 mm. The limiting factor when choosing the diameter is the current density in the winding, which for the degree of motor protection of IP54 should not be higher than 2.5 A/mm^2.

For the prototype, new packages of stator poles, a package of the rotor and stator coils were made, the shaft, bearing shields, bearings, labyrinth seals and other necessary parts were preserved from the 4PF200 DC motor in addition to the charge bed.

The switched reluctance motor SRM-14-615 is the totally enclosed air over (TEAO), 4 phase machine with 8/6 magnetic system topology. It has the next main data (table 1):

Table 1. Main data of SRM-14-615.

Parameter	Value
External stator diameter, mm	396.0
Rotor diameter, mm	261.0
Stator package length, mm	215.0
Air gap, mm	1.0
Coil turns number	23.5
Phase inductance, mH	29.0
Torque at I=I_n, Nm	5.5
Torque at I=2I_n, Nm	16.3

The second design stage it is calculating (using the QuickField program) the motor electromagnetic field for different current values at the agreed and mismatched positions of the rotor and the stator and also determining the flux linkage dependences on the phase current at different rotor positions.

3. Investigation results

Previously, the authors created the models in MATLAB/Simulink environment and investigated SRM with 6/4 and 12/8 magnetic system topology [4-5]. The Simulink model structure is the same for both topologies, so for creating Simulation SRM block diagram with 8/6 magnetic system topology it is necessary to add the fourth phase and to implement the differences by changing *Look-Up Table* blocks and the MATLAB *Function task* blocks.

The model of speed control developed in the MATLAB/Simulink environment with current limitation is shown in figure 1 where one phase is shown in detail, and the other three are represented by Subsystem blocks. The running motor speed value ω obtained after solving the equation of motion, is fed to the input of MATLAB *Function* 2 with *Integrator* 3.The specified parameters are: voltage U, current limit $I_{ref,}$ load torque T_{load}, moment of inertia J, switching on θ_{on} and off θ_{off} angles. The simulation determines the speed ω.

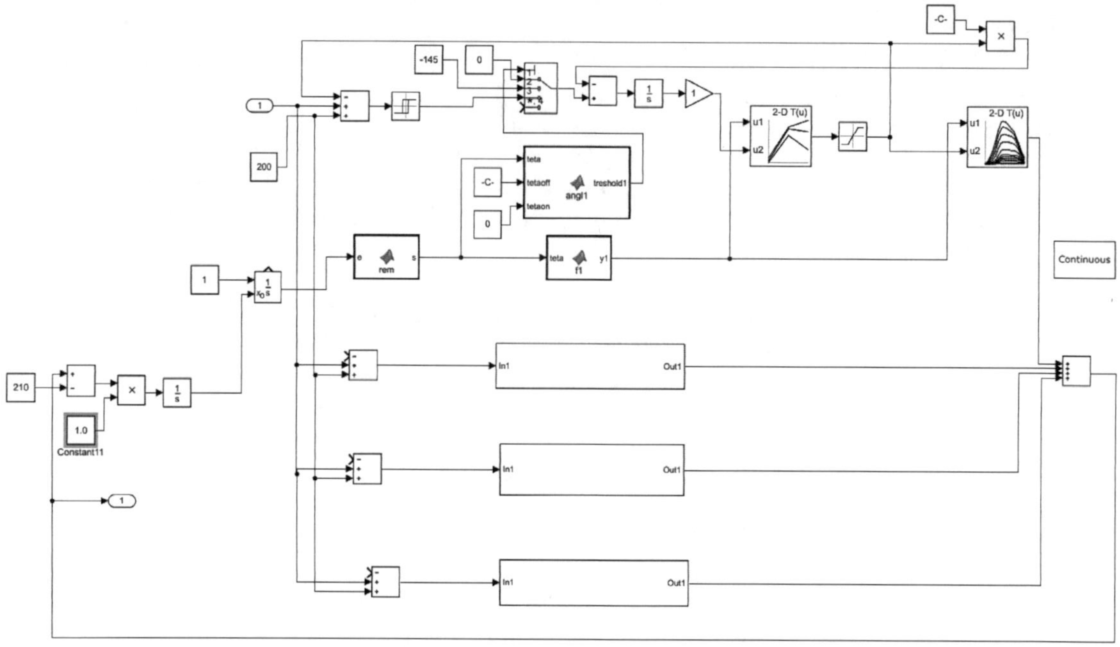

Figure 1. Simulation SRM block diagram for speed control with current limitation.

In the *Look-Up Table* block, the phase current data is recorded in the form of a 2-dimensional table, depending on the discrete values of flux linkage and misalignment angles $i=f(\psi,\theta)$. Flux linkage data obtained by finite element method (2-D FEM) in the form of a 2-dimensional table depending on the discrete values of phase currents and misalignment angles $\psi=f(i,\theta)$ have been converted into the dependences $i=f(\psi,\theta)$ by means of MATLAB by creating a special *m*-file. Current dependences on flux linkage at different rotor positions θ for SRM-14-615 are represented in figure 2.

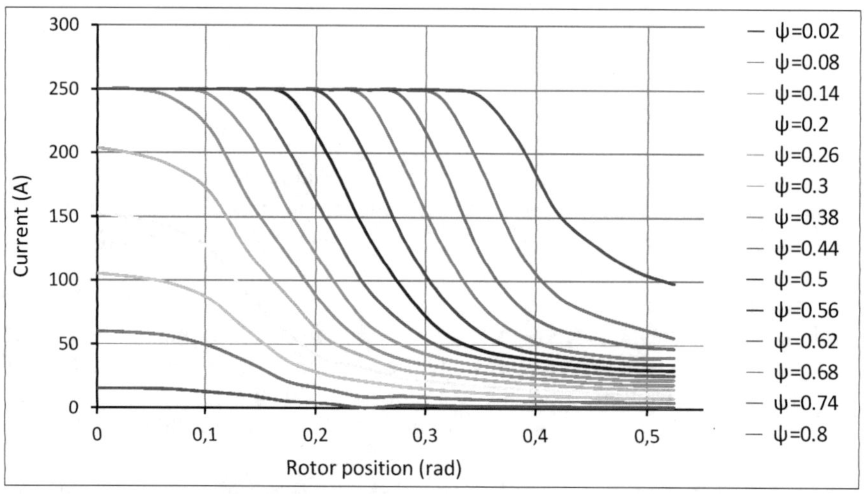

Figure 2. Current dependences on flux linkage at different rotor positions θ for SRM-14-615.

Flux linkages are obtained by two dimensional (2-D FEM) models for misalignment angles of 0–π/6° with π/90 interval. In the *Look-Up Table* 1 blocks, the torque data are recorded depending on the phase current and the misalignment angle. The problem is also solved by MATLAB *m*-file creating, in which the coenergy and its gradient have been determined. Torque dependences on currents at different rotor positions θ for SRM-14-615 are represented in figure 3.

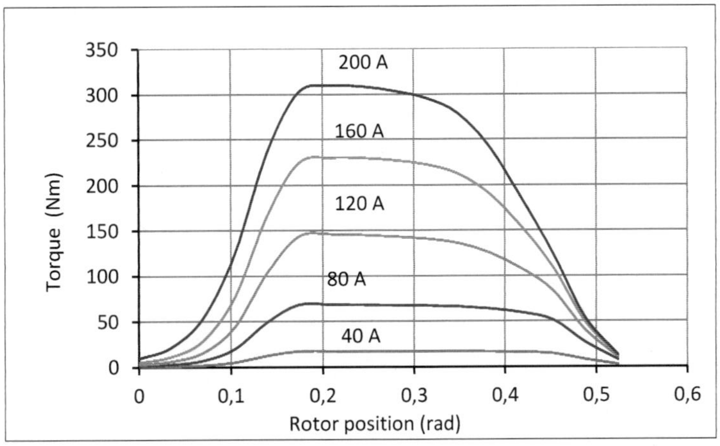

Figure 3. Torque dependences on currents at different rotor positions θ for SRM-14-615.

The MATLAB *Function* 2 block has to use the *rem* function (*remainder after division*) to convert a continuous rotor angle function into a *rotation angle* function between 0 to π/3 for topology 8/6. The MATLAB *Function* 7 block the created *m*-file is used. The rotation angle values for each phase are converted from the range [0-π/3] into the range [0–π/6] for topology 8/6., i.e. for the ranges for which the two-dimensional tables of the *Look-Up Table* and *Look-Up Table* 1 blocks are written.

The MATLAB *Function* block uses an *m*-file for each phase to switch signals in the multiport switch according to expression:

$$\begin{aligned} u_j &= U_{dc}, \theta_{on} \le \theta \le \theta_{off}, \\ u_j &= -U_{dc}, \theta_{off} \le \theta \le \theta_{ent}, \\ u_j &= 0, \theta \ge \theta_{ent}, \end{aligned} \tag{1}$$

where θ is the misalignment angle of the rotor and stator position, u_j is the phase voltage depending on the angle θ.

For the further analysis, phase current, flux linkage, and torque data are stored in the workspace using *Simout* blocks. Figure 4 contains the example of data received on the *Scope oscilloscope* screens during simulation.

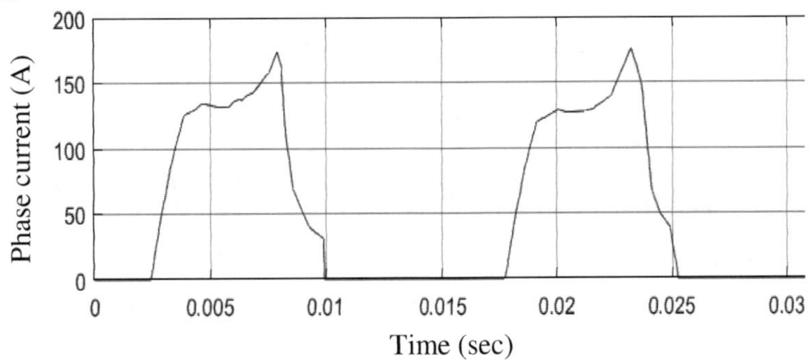

(a)

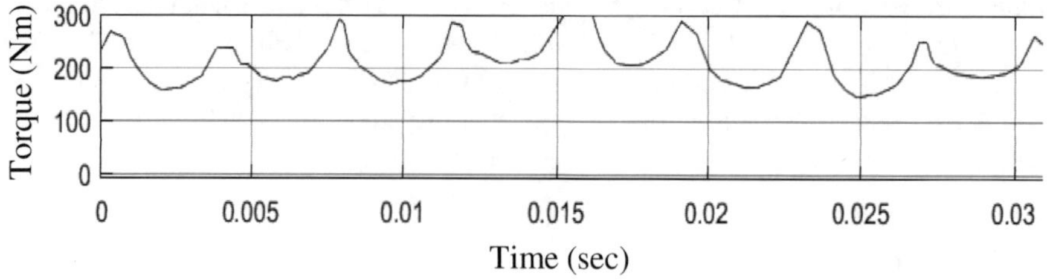

(b)

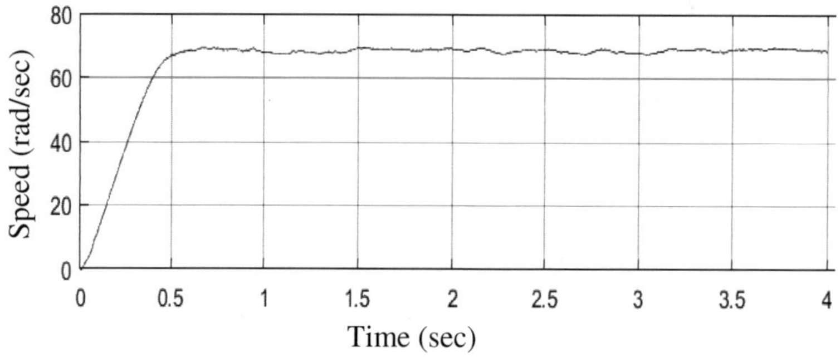

(c)

Figure 4. Simulation data on screens oscilloscopes (mode $\Delta=\pi/8$, I_{rms}=71 A, T_{el}=215.6 Nm, ω=68.6 rad/s): phase current (a), electromagnetic torque (b), motor speed (c).

The mine battery electric locomotive AM8D is equipped with two DC motors DPT-14, which can be schematically switched on in parallel or in series. The simulation was performed for a voltage of 145 and 72.5 V to take into account both switching options. The results are presented in table 2 and figure 5.

Table 2. SRM-14-615 motor speed simulation on the reference model.

I_{ref}, A	n, rpm	I_{rms}, A	T_{el}, Nm
	U=145.0 V		
200	1108	27.7	32.9
200	927	36.7	50.5
200	701	53.8	127.4
200	655	71.1	215.6
	U=72.5 V		
200	676	25.5	21.4
200	575	29.6	28.7
200	473	39.1	48.7
200	386	44.6	100.3

Figure 5 also shows the calculated points of the mechanical characteristics of the DRT-14 motor. The comparison shows that the designed motor SRM-14-615 meets the required operating modes as part of the AM8D mine battery electric locomotive.

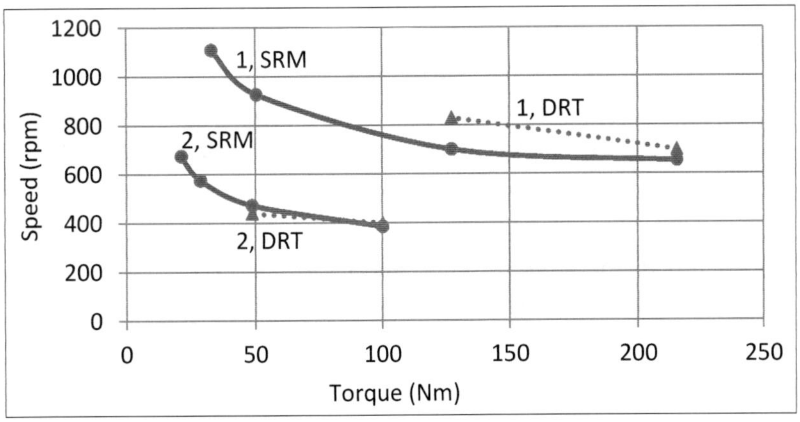

Figure 5. SRM-14-615 motor speed simulation on the reference model: 1 - U=145.0 V, 2 - U=72.5 V.

It is possible to study the switched reluctance motors of using artificial intelligence methods and techniques [7]-[10].

The basis for such development SRM-14-615 with the magnetic system topology 8/6 is the new created model in the MATLAB/Simulink environment for speed control with current limitation (figure 1).

A model with a fuzzy control system is developed in the Fuzzy Logic Toolbox package, and a model with a neural network control system is developed in the Neural Network Toolbox package. First of all in Simulation SRM block diagram for speed control with current limitation (figure 1) has been added fragment figure 6, a to create simulation SRM block diagram for speed control with a fuzzy logic, or fragment figure 6, b to create simulation SRM block diagram for speed control with the neural network. The connection points of the fragments are shown in figure 1 and figure 6.

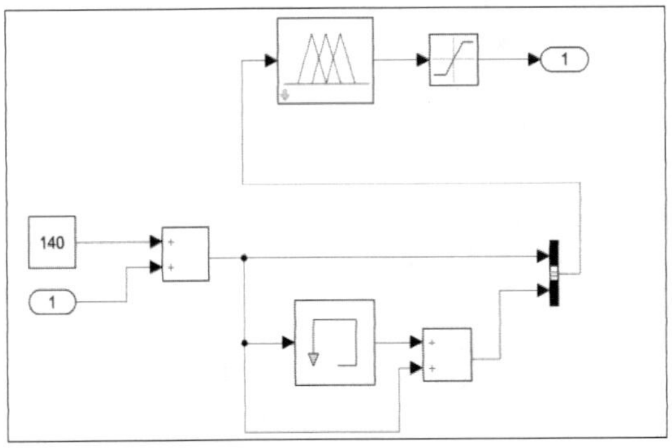

(a)

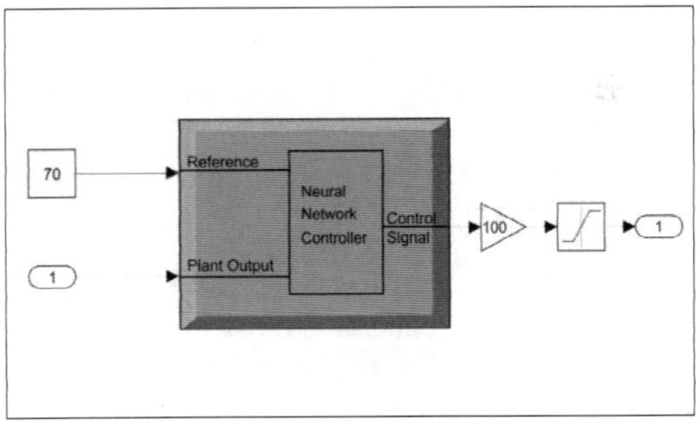

(b)

Figure 6. Fragments for creating simulation SRM block diagram for speed control with a fuzzy logic (a) and with the neural network (b).

The model with a fuzzy logic-based control system is developed interactively in the Fuzzy Logic Toolbox package. A fuzzy inference system of the Sugeno type has two input variables – the angular rotation frequency ω and its deviation from the specified one $\Delta\omega=\omega-\omega_{ref}$ is selected.

Each variable is characterized by 5 membership functions, the type of which can be selected from the functions built into the editor. For both variables, triangular functions of the type *trimf* are selected (S2, S1, CE, B1, B2). For membership functions 25 rules of the fuzzy inference system are written in the form: "If (x1 is S2) and (x2 is S2) then (y is S2)". The adopted rules are presented in table 3.

Table 3. 25 rules of the fuzzy inference system.

ω / Δω	S2	S1	CE	B1	B2
S2	S2	S2	S2	S1	CE
S1	S2	S1	S1	CE	B1
CE	S2	S1	CE	B1	B2
B1	S1	CE	B1	B1	B2
B2	CE	B1	B2	B2	B2

The method of algebraic product (*prod*) is chosen for performing logical conjunction under fuzzy rules, and the method of algebraic sum (*probor*) is chosen for performing logical disjunction. To make a logical conclusion in each of the fuzzy rules, the minimum value (*min*) method is selected, and for aggregating values, the maximum value (*max*) method is selected. To perform de-fuzzification of output variables in the Sugeno fuzzy inference system, the weighted average method (*wtaver*) is adopted. According to the selected conditions, the *srm_nFL_86* file is created, which is written to the Fuzzy logic controller with Ruleviever Simulink block of the model shown in figure 6, (a).

The model with the neural network control system is developed in the Neural Network Toolbox package. The controller based on the reference model – Model References Controller - is selected as the control system (figure 6, b). The Simulink file *srm_n_*86 of motor speed control by current limitation is selected as the reference model of the SRM-14-615 for the neural network (figure 1). As a training example, the *base*86.mat file is created, which records the dependence of the angular speed of rotation on the time ω=*f*(*t*), obtained by calculating the reference model *srm_nNN_*86 for the speed of rotation ω_{ref}=68.6 rad/s and containing 10620 values.

The Model References Control window also permits to select data: the number of hidden layers – 10, delayed inputs of the reference model – 2, delayed outputs of the controller – 1, delayed outputs of the object – 2.

To identify the object model (*Plant indetification*), the following parameters are set: the number of hidden layers – 10, delayed inputs of the object model – 2, delayed outputs of the object model – 2; the Simulink file *srm_nNN_86* which contains the controller (figure 6, b), and the training example file *base86.mat* are entered; the number of variables (10620), minimum (0) and maximum (ω_{ref}) values, maximum (0.001) and minimum (0.0001) interval values. When training the network for the specified number of iterations and training the controller, we get the final acceptance of the Simulink data and then the file *srm_nNN_*86 is ready for modeling.

Results of SRM-14-615 motor speed simulation on the reference model for the mode with the voltages U=145°V and 72.5°V are shown in figure 5. The same modes are modeled on the created models. For comparison, the data are summarized in table 4.

Table 4. Comparison of SRM-14-615 simulation results for different control algorithms

Reference model				Model with a fuzzy logic, figure 6, a				Model with the neural network figure 6, b			
I_{rms}, A	Ω, rad/s	T_{mean}, Nm	K_{pT}, %	I_{rms}, A	ω, rad/s	T_{mean}, Nm	K_{pT}, %	I_{rms}, A	ω, rad/s	T_{mean}, Nm	K_{pT}, %
					U=145.0 V						
71.1	68.6	215.6	25.6	81.7	68.6	213.7	24	80.4	68.4	221.4	23.4
31.1	116	32.9	46.5	34.5	100.4	44.8	42.4	37.5	114.1	32.6	48.2
					U=72.5 V						
44.6	40.4	100.3	27.8	53.1	40.9	99.25	29.4	53.3	40.3	98.1	32.4
25.5	70.7	21.4	48.6	26.5	71.0	21.3	48.5	27.4	69.7	24.3	47.6

The comparison was carried out according to the main parameters of the studied mode – the effective value of the phase current I_{rms}, the angular rotation frequency ω, the average value of the rotating electromagnetic torque T_{mean}, the torque ripple coefficient K_{pT}. It should be noted, that there is a minimal difference in the angular frequency of rotation (1.6%), a small difference in the torque magnitude (3.6%) and its ripple coefficient, and the largest difference in the magnitude of the RMS current (3.9–19.5%).

4. Conclusion

Switched reluctance motor SRM-14-615 based on the principles of block-modular motor design is proposed for electric drive of mine battery electric locomotives instead of the DRT-14 DC motor. The study of the SRM-14-615 operating modes in MATLAB/Simulink environment was conducted on a model created for the 8/6 magnetic system configuration. The model makes it possible to improve the design procedure, rationally select the control parameters (I_{ref}, θ_{on}, θ_{off}, ω, M_{load}) and get the dynamics of the main parameters changes in different motor modes. SRM-14-615 has the mechanical characteristics as DC motor DRT-14 and meets the required operating modes as part of the AM8D mine battery electric locomotive.

Also two types of models using methods and techniques of the artificial intelligence theory are presented: a model with a fuzzy control system in the Fuzzy Logic Toolbox package and a model with a neural network control system in the Neural Network Toolbox package. Based on the results of verification of the created models in various modes it has been found that the selected methods and characteristics of artificial intelligence controllers allow us to successfully simulate the Switched Reluctance drive. As in Fuzzy Logic Toolbox package, so in Neural Network Toolbox package only one training example file can be used for a wide range of motor speed.

Acknowledgements
The work has been carried out with the financial support of the RFBR grant, project No. 20-08-00386.

References

[1] Miller T J E 2002 *IEEE Trans. on Industrial Electronics* **49(1)** 15-27
[2] Pratapgiri S, Polaki P and Narsimha V 2012 *World J. of Modelling and Simulation* **8(3)** 185-193
[3] Krishnan R 2001 *Switched reluctance motor drive* (CRC Press)
[4] Antipov V N, Grozov A D and Ivanova A V 2016 *Elektrichestvo* **10** 45–52
[5] Antipov V N, Grozov A D and Ivanova A V 2020 *Proc. of the Int. Conf. Electr. Mach. ICEM 2020* 1820–1825
[6] Chancharoensook P and Rahman M F 2002 *Proc. IEEE Annual Conf. of the Industrial ElectronicsSociety*, **1** 491-496
[7] Tanaka K and Wang H O 2001 *Fuzzy Control Systems Design and Analysis: A Linear Matrix Inequality Approach* (New York: John Wiley & Sons, Inc.)
[8] Wadnerkar V S, Bhaskar M M, Das T R and RajKumar A D 2010 *J. of Electrical Engineering & Technology* **5(2)** 276–281
[9] Pushparajesh V, Balamurugan M and Ramaiah N 2019 *SSRN Electronic J.* DOI: 10.2139/ssrn.3371369
[10] Li C, Wang G, Li Y and Xu A 2018 *Automatika* **59(1)** 24–34

Study of the thermal characteristics of the aluminum radiator ROYAL Thermo Evolution

Z G Mar'ina[1], A Y Vereshchagin[1], A V Novozhilova[1], M A Komarevtsev[1] and K O Isaeva[1]

[1] Northern Arctic federal university named after M. V. Lomonosov , 17, Northern Dvina Emb., Arkhangel'sk, Russia

Abstract. The use of aluminum radiators in heating systems began with the foreign companies products such as Fondital Group, Global Radiatori, Ferroli, etc. There are now Russian aluminum radiators companies that adhere to European standards, for example ROYAL Thermo, Rifar. The thermal characteristics of radiators are usually specified by the manufacturer. However, they are not always confirmed in practice. This is due to the fact that radiator connecting method is not taken into account. In some technical data sheets, the power reduction factors are given depending on the difference between the average coolant temperature and the air temperature in the room. Therefore, the study of the influence of aluminium heating appliances connecting on its thermal characteristics is an actual task. The article presents the results of research of radiator of the ROYAL Thermo Evolution type, it was made on request from company specialising in heating systems design. It was found that the actual thermal power of one section varies in the range of 135 ... 225 W, depending on the connection method of the radiator, the thermal power of the section declared by the manufacturer corresponds only to the diagonal «top-bottom» connection, the thermal power of the eight-section radiator with one-sided «top-bottom» connection is 12% higher than with a diagonal one. Studies show, when designing heating systems, it is necessary to take into account the connection method and the number of device sections.

1. Introduction

A large variety of heating equipment nowadays has led to the fact that the thermal performance of radiators is usually specified by the manufacturer. However, it is not always confirmed in practice. The review of numerous sources on this issue allows us to conclude that heating studies confirm data only for certain types of radiators [1–10].

There are a large number of modern heating devices on the Russian market nowadays and a large proportion is accounted for by aluminum radiators. Their undoubted advantage is high heat transfer, which stands at 200÷210 W, while the volume of each radiator section is only 450 ml. They have an aesthetic appearance, what allows to "introduce" such radiators into the interior design of any style. The low inertia of aluminum heating devices allows to regulate the temperature regime precisely with the help of thermostatic devices. These advantages, as well as a long service life and a large variety of design possibilities and size types of aluminum radiators led to their widespread use in Russia and abroad.

An eight-section aluminum radiator of the ROYAL Thermo Evolution type intended for heating systems in residential, public and industrial buildings was presented for the research. The technical specifications of the device are presented in table.

Table. The technical specifications of the ROYAL Thermo Evolution type radiator.

Working pressure, MPa	Heat transfer of one section, W	Maximum temperature of the coolant, °C	Capacity of one section, l	Centre-to-centre dimension, mm	Weight, kg	Life cycle, years
1.6	203	110	0.35	500	1.3	25

To date, the Government of the Russian Federation has adopted the Resolution No. 717 of June 17, 2017 on the introduction of an obligatory certification system of all types of heating devices [11], their accordance with State standard [12]. Thus, the deviations of the nominal heat flux claimed by the manufacturer from the indicators established by the test results, should not exceed the maximum permissible tolerances (from – 4 % to + 5 %) [12]. It is of interest to determine the actual heat flux of a heating device, therefore the aim of the work is to study the thermal characteristics of the heating device of the ROYAL Thermo Evolution type [13].

At present, when designing heating systems, the manufacturer does not give recommendations about the influence of the device connecting method on the change in heat output, therefore, the influence of the device connecting method on its heat transfer was also determined.

2. The Scheme of the Experimental Installation

The studies were conducted in the laboratory of the training-scientific center of energy innovations of the Heat Technology and Heat Power Engineering Department NArFU.

The company presented an eight-section aluminum radiator of the ROYAL Thermo Evolution type for the research, (figure 1).

The increased thermal power of the radiator is provided not only by using aluminum as a high thermal conductive material, but also by the design of the fins. The fins form vertical rectangular channels in which air moves freely from bottom to top, increasing the heat transfer of the device during natural convection. Smooth transition of the heat flux from the fin spacing provides direct air flow into the room.

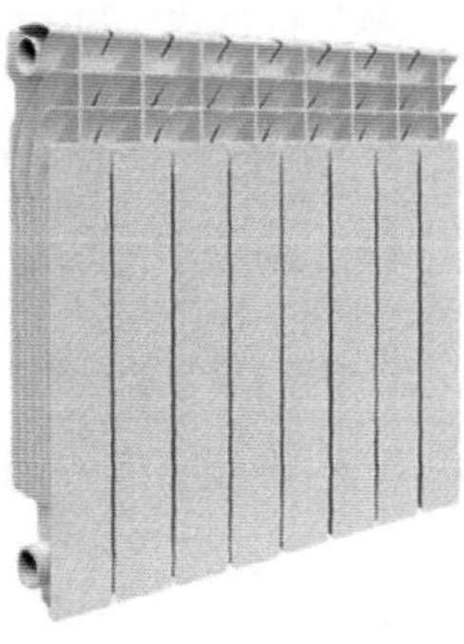

Figure 1. The appearance of the radiator.

The scheme of the experimental installation is shown in figure 2. The water heater 2 made it possible to obtain a coolant temperature up to 80 °C. The circulation of the coolant was provided by the pump 3. The heat load of the device was determined by the indications of the flow meter (4) and two Chromel-Copel thermocouples (9, 10). Thermocouples are caulked in metal inlets to the device at the inlet and outlet from it and connected via a multi-position switch 7 to a millivoltmeter 8. The cold junctions of the thermocouple are immersed in the Dewar vessel 6 with melting ice. A screen filter 5 [14–15] is installed in front of the flow meter for the pollution capture.

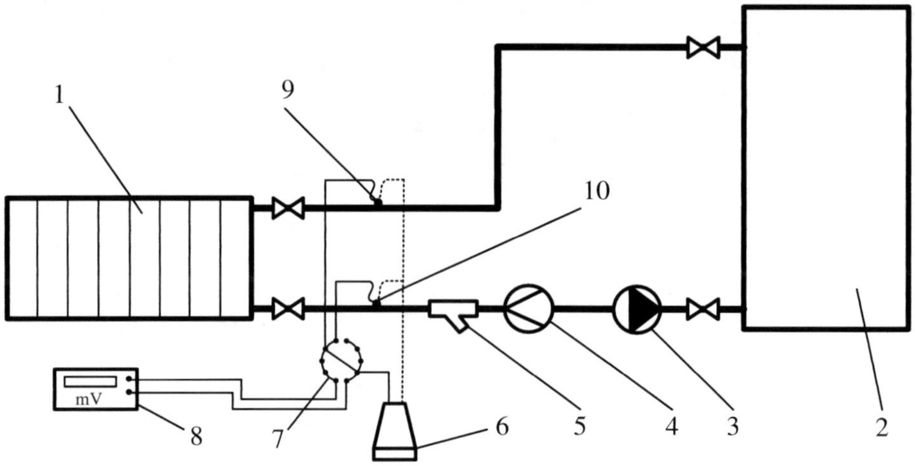

Figure 2. The scheme of the experimental installation.

By the stationary thermal mode, there were made measurements of the coolant flow rate, the emf. of the thermocouples installed at the inlet and outlet of the radiator, and the air temperature in the room. The experiments were carried out in the range of temperatures of the heating coolant 40...80 °C at flow rates of 0.13...0.22 m^3/h. The thermal load of the heater was determined by the flow rate and temperature difference of the coolant. The uniformity of the temperature field on the outer surface of the fins and the direction of the coolant's movement in the device was determined using a TESTO-350 thermal imager.

3. The study of the heating characteristics of aluminum radiator

Let us determine the thermal power of a radiator ROYAL Thermo Evolution with different methods of connection.

In the equipment certificate the thermal power of the section, claimed by the manufacturer, is $q_n = 203$ W with the calculated difference between the average coolant temperature and the air temperature in the room $\Delta t_{av} = 70$ °C, obtained at coolant temperature 110/70 °C and air temperature 20 °C [4]. In Russia, two-pipe heating systems are designed at coolant temperatures before and after the heating device of 95 °C and 70 °C respectively, and the air temperature in the room is 20 °C, which corresponds to the calculated temperature difference $\Delta t_{av} = 62.5$ °C.

In other design conditions, manufacturers offer to use the correction factor $k = f(\Delta t_{av})$ and determine the density of the heat flux section by the formula:

$$q = q_n \cdot k. \quad (1)$$

The thermal power of the radiator at different coolant temperatures and methods of connection is shown in figure 3.

The change in consumption does not affect the heat transfer of the radiator with one-sided and diagonal «top-bottom» connections (lines 1, 2). Impact of costs should be taken into account with «bottom-bottom» connection (lines 3).

The heat flux density of the eight-section radiator declared by the manufacturer, taking into account the influence of Δt_{av}, calculated by formula 1, is shown by line 4.

Comparison of the declared and experimental data shows that with «top-bottom» connection there is a stock of thermal power, and with «bottom-bottom» connection, the number of sections should be determined with caution, because the device power will be lower than declared by the manufacturer and significantly depend on the coolant flow rate.

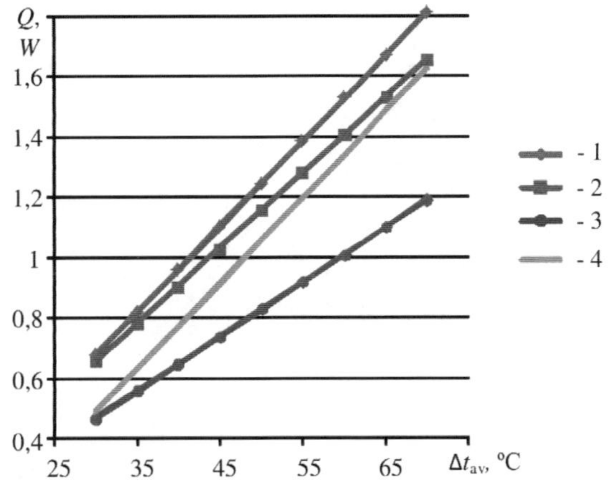

Figure 3. Thermal power of radiator.

To study the direction of the coolant in the device with a one-sided «top-bottom» connection, a TESO-350 instrument was used for thermal imaging from the beginning of the coolant supply until the radiator was fully heated. Thermal imaging was performed at intervals of 20 seconds. The 75 °C coolant enters the cooled radiator from the top, heats the upper collector, and then the vertical columns (figure 4).

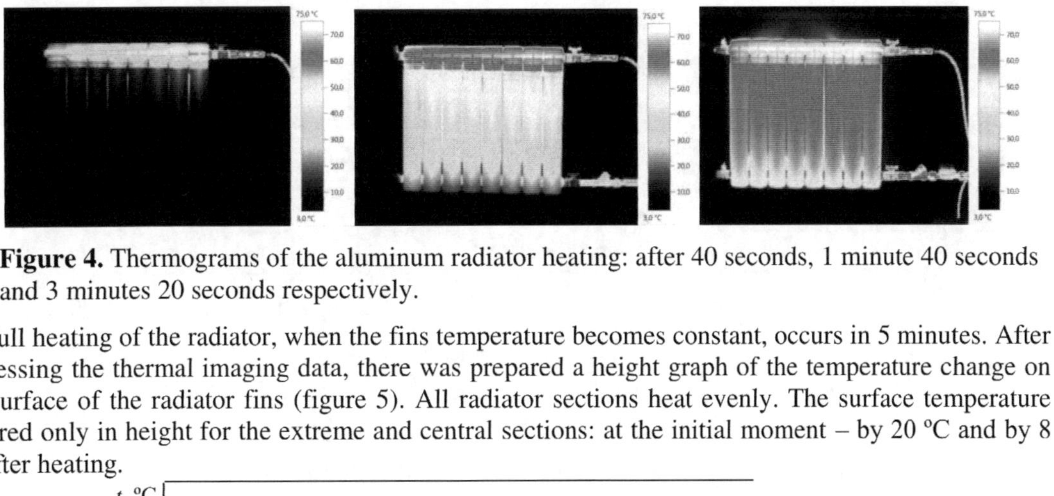

Figure 4. Thermograms of the aluminum radiator heating: after 40 seconds, 1 minute 40 seconds and 3 minutes 20 seconds respectively.

Full heating of the radiator, when the fins temperature becomes constant, occurs in 5 minutes. After processing the thermal imaging data, there was prepared a height graph of the temperature change on the surface of the radiator fins (figure 5). All radiator sections heat evenly. The surface temperature differed only in height for the extreme and central sections: at the initial moment – by 20 °C and by 8 °C after heating.

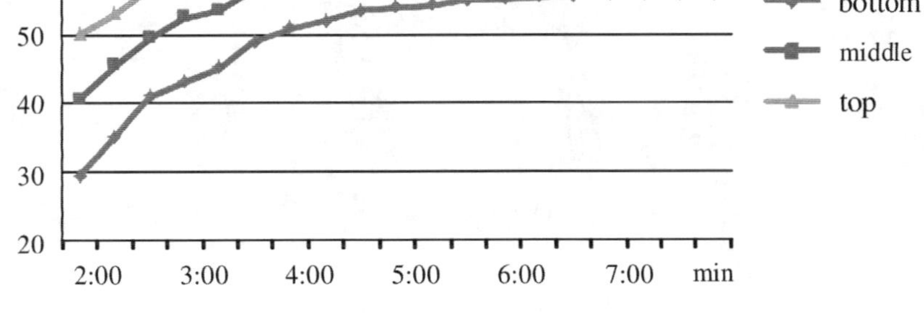

Figure 5. Fins surface temperature of the sections during the radiator heating.

The temperature difference between the coolant and the fins surface after heating was on average 13 °C.

It can be seen that with diagonal «top-bottom» connection of the radiator, the coolant fills the heating device through the furthest (eighth) section with a gradually heating from the last section to the first (figure 6).

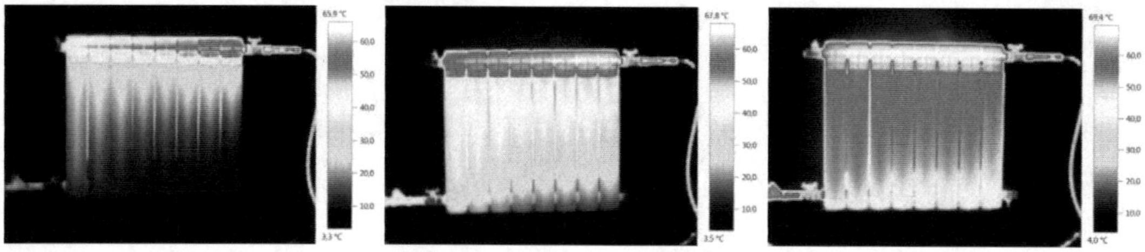

Figure 6. Thermograms of the aluminum radiator heating: after 1 minute, 2 minutes and 5 minutes, respectively.

Based on the analysis of thermograms (figure 7), it can be seen that the heating of the eighth section proceeds more evenly, as the coolant circulation begins to flow through it. The temperature difference between the top and bottom of the fin is on average 5 °C. The first section starts to heat with a delay. The difference between the upper and lower points of the fin decreases from 20 °C at the beginning of heating to 10 °C. This difference can be explained by a different amount of coolant flow through each section, as evidenced by the lower thermal power of the radiator (by 12%) compared to a one-way connection.

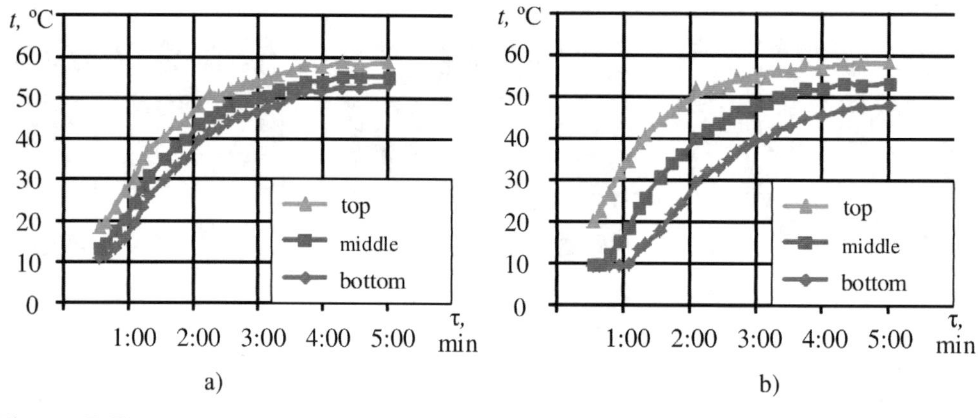

Figure. 7. The temperature of the fins' surface of the aluminum radiator sections: eighth section (a) and first (b).

Figure 8 shows the thermogram of the radiator heating with «bottom-bottom» connection. In this case, the radiator heating is as follows. The coolant rises through the first section to the upper collector and then moves through the other sections from top to bottom.

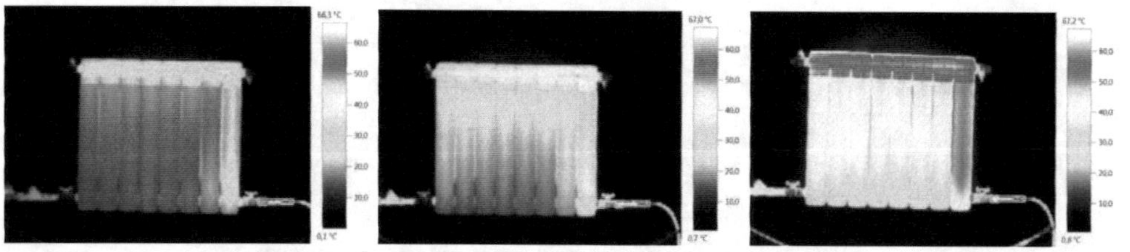

Figure 8. Thermograms of the aluminum radiator heating: after 1 minute, 2 minutes and 5 minutes, respectively, with «bottom-bottom» connection.

4. Conclusions

1) The thermal power of the section declared by the manufacturer is 203 W, it corresponds only to the diagonal «top-bottom» connection.
2) The thermal power of the device in the design conditions with one-sided connection is 12% higher than with a diagonal one.
3) The actual thermal power of the section varies in the range of 135 ... 225 W at 70 ºC, depending on the connection method of the radiator. The discrepancy is -33...+11%.
4) With «bottom-bottom» connection method it is necessary to consider the effect of coolant flow.

References

[1] https://otivent.com/sravnenie-radiatorov-otopleniya-po-teplootdache
[2] I. Plokhikh. *Aqua-Term.* **2** (2011) [Electronic resource]. – Access mode: https://aqua-therm.ru/articles/articles_185.html.
[3] M. Embaye, R.K. Al-Dadah, S. Mahmoud Energy and Buildings **121**, 298-308. (2016)
[4] L. Brady, M. Abdellatif, J. Cullen, J. Maddocks, A. Al-Shamma'a Energy and Buildings **133**, 414-422. (2016)
[5] M. Maivel, M. Konzelmann, J. Kurnitski Energy and Buildings **86**, 745-753 (2015)
[6] A. Gheibi, A.R. Rahmati Applied Thermal Engineering **163**, 25 (2019)
[7] R. Marchesi, F.Rinaldi, C. Tarini, F. Arpino, G. Cortellessa, M. Dell'Isola, G. Ficco Thermal Science **23,** Part B 989 – 1002 (2019)
[8] Y. Jian, Z. Yu, Z. Liu, Y. Li, R. Li *Procedia Engineering* **146**, 466 – 472 (2016)
[9] H.G. Hameed, H.A.N. Diabil, M.M.A. Saeed *Journal of Mechanical Engineering Research and Developments* **44** (5), 420 – 432 (2021)
[10] D. Laouali, F. Beniere *Journal of Materials and Environmental Science* **3** (1), 34 – 49 (2012)
[11] http://www.consultant.ru/cons/cgi/online.cgi?req=doc&base=LAW&n=218692&fld=134&dst=1000000001,0&rnd=0.8020253409431399#018911350397033821
[12] State standard No 31311-2005 «Heating devices. General specifications». M., Standartinform (2006)
[13] http://www.royal-thermo.ru/catalog/radiatory_otopleniya/alyuminievye_radiatory/
[14] A.Yu. Vereshchagin, N.V. Latishova *Lomonosov scientific readings of students, postgraduates and young scientists – 2017* (collection of conference materials), 400-404 (2017)
[15] N.V. Latishova, A.Yu. Vereshchagin *Lomonosov scientific readings of students, postgraduates and young scientists – 2017* (collection of conference materials), 1080-1084 (2017)

Boilers slagging when burning wood pellets

V K Lyubov, A V Malkov and P D Alekseev

Northern (Arctic) Federal University named after M.V. Lomonosov,
Russia, 163002 Arkhangelsk, Severnaya Dvina emb., 17

E-mail: v.lubov@narfu.ru

Abstract. A promising trend for upgrading wastes from timber cutting, processing and treatment is their granulation. It allows to increase their specific heats of combustion by 2.5–3.5 times and their portability characteristics by 3–4 times, to reduce transportation costs by 6–10 times and to improve all the operations stages. The construction and commissioning of boiler facilities operating on refined biofuel made it possible to form a stable domestic market for wood pellets. However, 0.5 – 1.5 MW nominal capacity hot water boilers equipped with furnaces and profiled burners at the bottom, in cold seasons had fast accumulation of focal residues deposits in the burners and on the furnace chambers lining. The process was complicated by these deposits hardening due to their melting and sintering. These circumstances cause a decrease in the energy and environmental performance of heat-generating installations and their reliability, and also leads to the unplanned shutdowns to clean the boiler furnaces. To find out the reasons for these negative phenomena and to develop recommendations for their elimination, a set of research operations was carried out with wood pellets shipped by the manufacturer and supplied to the burners of the boilers under the analyses; with focal residues accumulated in the burners and on the lining of the furnace chambers; as well as an analysis of the heat generating facilities operation modes. The studies carried out made it possible to identify the main factors that caused these negative phenomena and to develop the recommendations for their elimination.

1. Introduction

Energy supply problems are vitally important for any country in the world. As civilization develops, the specific energy consumption is constantly increasing. However, the traditional energy reserves are very limited, and they are constantly becoming more expensive. Meanwhile, peoples' activity is accompanied by large amount of wastes accumulation: wood, agricultural, solid household, sewage sludge, etc., the volumes of which are growing rapidly and pose a serious threat to the modern world.

Timber wastes accumulated from cutting, processing and treatment are classified as difficult to burn fuels due to their high moisture content, low energy density and extremely heterogeneous particle size. A promising approach for such wastes upgrading is their granulation [1]. Since the beginning of the XXI century, this industry has been intensively developing in Russia. The country bioenergy potential, as well as fossil hydrocarbons one, surpasses those of any region in the world [1-3].

Considering the fact that towns energy and environmental problems are closely interconnected and have to be addressed in a comprehensive manner, simultaneously and by utilizig the resource-saving technologies, a program was developed in the Arkhangelsk region to replace imported fuels with biofuels, to upgrade the existing energy facilities and to build the new ones: modular, fully automated, operating on refined environmentally friendly biofuel - granules. So, since 2010, more than 100 regional energy facilities have been commissioned, upgraded and transferred to the use of solid biofuels.

As an example, the total installed capacity of six new fully automated boiler houses of LLC "Arkhbioenergo" operating on wood pellets amounted to 46 MW, and the achieved gross efficiency of hot water boilers is more than 90.0%. At the same time, carbon monoxide emissions, soot particles and nitrogen oxides were minimal [4]. The achieved gross efficiency of biofuel boilers corresponds to the efficiency of fuel oil boilers in good technical condition. The construction and commissioning of boiler houses operating on refined biofuel made it possible to form a stable domestic market for wood pellets. This circumstance provided additional internal guarantees for the dynamic work of pelleted fuel regional producers.

However, 0.5 – 1.5 MW nominal capacity hot water boilers equipped with furnaces and profiled burners at the bottom had (in cold seasons) fast accumulation of focal residues deposits in the burners and on the furnace chambers lining. The process was complicated by these deposits hardening due to their melting and sintering. These circumstances caused a decrease in the energy and environmental performance of heat-generating installations and their reliability, and also led to the unplanned shutdowns to clean the boiler furnaces.

The focal residues processes at high temperatures (softening, melting, fluidity of the resulting melt) present one of the most important technological characteristics of energy fuel. These characteristics significantly affect the boiler design, its reliability, as well as the operating costs [5, 6].

Granular fuel in such hot water boilers is fed by auger feeders to the burners, the primary air passes through the gaps of the grates made of refractory cast iron with chromium addition and penetrates the fuel layer. To ensure the afterburning of the fuel combustible components, secondary air is supplied to the above-layer zone of the burner through nozzles located oppositely on the side walls.

To clarify the reasons for the above listed negative phenomena and to develop the recommendations for their elimination, the research work was carried out with wood pellets entering the burners, with focal residues formed in the burners and on the furnace chambers lining, as well as an analysis was performed for the energy facilities operating modes.

2. Objects, materials and methods of research

The determination of the elements composition of wood pellets and raw materials used for their production, as well as of focal residues, was carried out using an automated Euro EA-3000 analyzer according to the CHNS analysis standard (carbon, hydrogen, nitrogen and sulfur). In addition, an EDX-8000 X-ray fluorescence spectrometer was used to determine the elements composition of the materials under study. The device allows simultaneously detect, measure and record the radiation intensity of various elements. The studies were carried out at the Scientific Equipment Center of the Scientific Association "Arctic" in the Northern (Arctic) Federal University (NArFU) named after M.V. Lomonosov in Arkhangelsk city.

Thermal analyzes of wood fuels and focal residues were carried out using an IKA C 2000 basic Version 2 calorimeter with a LOIP FT-216-25 liquid cryothermostat and with the units of the complex thermal analysis laboratory of the Educational and Scientific Center for Energy Innovation of NArFU. Investigations of the granulometric composition of the fuel and focal residues were carried out by the sieve method using the Retzsch AS 200 Control and "029" analyzers. The portable NHP 100 tester from the German company Holmen was used to study the mechanical strength and abrasion of the granules. minimum hot water boilers heating capacity is 20% of the nominal.

The units under the research were the automated power sources equipped by Hekotek with 0.5–1.5 MW nominal capacity hot water boilers. The vertical cylinder tanks were mounted next to the boiler houses to provide the granular fuel stock for operations. The working pressure of hot water at the outlet of these heat-generating installations should not exceed 0.4 MPa, and the temperature should not exceed 115 °C. Hot water boilers have two circuits, with chemically purified deaerated water circulating in the first one. 110/90 °C is the calculated temperature mode for the primary circuit working conditions. Combustion products provide heating of the primary circuit boiler water by making three strokes in the boiler smoke tubes, then they are cleaned in a multicyclone. Heating of the secondary circuit water (70/90 °C) is carried out using a plate heat exchanger. Each boiler is equipped with an individual heat exchanger and the smoke pipe. To ensure smooth adjustments of the heat-generating installations performance, the drive of smoke exhausters, augers, fans of primary and secondary air has frequency control. The minimum hot water boilers heating capacity is 20% of the nominal.

The surveyed substances were sampled in accordance with the current requirements [7].

3. Results

Based on the analysis of the experience of burning granular fuel in the Russian Federation and abroad [1, 4, 8] the following preliminary conclusions were made:

1. The process of sintering and the agglomerates formation occurs when pellets contain small granules less than 3.15 mm in size and a large amount of dust particles. Such a finely dispersed mass has an increased bulk density and, upon entering the combustion zone, forms fuel agglomerates that are poorly blown by air flows; in these places, a crust begins to form, gradually increasing in size. During a long stay in the burner device, the combustible substances are almost completely oxidized. One of such ash and slag agglomerates, extracted from 1.5 MW nominal power hot water boiler, is shown in Fig. 1. There are several causes for such "small parts" formation in wood pellets:

- Low-quality pellets from the supplier, as a result of the plant’s equipment failure or pellets abrasion and destruction during transportation and reloading.
- Improper storage of pellets, namely: placing "big bags" on the ground without a stand (mat) that blocks wetting the pellets from the ground; or high humidity in the storage room, leading to crushing of the pellets in the lower third of the bag and literally turning them into dust.
- Incorrectly designed wood pellets storage area and boilers burners fuel supply system resulting in the pellets significant part been shredded.

2. Different impurities present in the wood granules, such as bark and sand, give another cause of sintering parts. Sand softens and melts in the high temperature zone, especially in a reducing atmosphere, which can also lead to slagging.

3. Boilers operations with significant deviations from optimal modes. This contributes to the creation of conditions for the initial formation of fuel and ash and slag agglomerates and the adhesion of molten mineral particles to the surface of the grate and to the boilers lining.

4. Boiler burners design without continuous stable withdrawal and tedding of focal residues formed in pellets burning process.

Figure 1. Sintered ash and slag agglomerate.

CJSC "Sawmill № 25" is the largest producer of pelleted fuel in the Arkhangelsk region. Pellets for the power facilities under analyses are supplied from its “Tsiglomensky” site, where the wood pellet preparation shop No.2 is located. An extended analysis of the granular fuel produced in the shop No.2 showed that the granular fuel sample meets the requirements of Russian and European standards in terms of all heat engineering and strength parameters, as well as in bulk density [9]. The main indicators of wood pellets are presented in Table 1.

Table 1. The main performance of the boiler.

Indicator, units of measurement	Working weight	Dry weight	Combustible mass
Moisture, %	5.74		
Ash content, %	0.52	0.55	
Combustion heat lowest, MJ / kg	17.615		18.943
Combustion heat, highest, MJ / kg			20.422
Sulfur, %	0	0	0
Hydrogen, %	6.12	6.49	6.53
Nitrogen, %	0.19	0.20	0.20
Chlorine, %	-	-	
Lead, %	-	-	
Vanadium, %	-	-	
Molybdenum, %	-	-	
Volatiles, %	78.78	83.58	84.04
Mechanical strength, %	98.45		
Average diameter, mm	8.06		
Average apparent density, g / cm^3	1.18		
Bulk density, g / cm^3	0.695		
Coke residue		Powdery	

4. Results

The study of the wood pellets particle size distribution showed: the weight content of chips (particles with a size of 3.15 mm or less) is much less than 1.0% (Fig. 2); granules with a length of 40 mm or more are absent, which meets Russian and European standards [10]. The performed analyzes also showed: granules of 12 to 25 mm length dominate in the fuel under study; their mass fraction was 83.52% (polydispersity coefficient n = 5.618, and the coefficient characterizing the composition size b = $1.15 \cdot 10^{-24}$).

Based on the above, there are no claims to the quality indicators of the wood pellets manufacturer.

The study of the particle size distribution, strength and abrasion of wood pellets taken from the auger feeder in front of the boiler with 1.5 MW rated power showed that the pellets have a much less homogeneous and finer composition (n = 2.369; b = $1.619 \cdot 10^{-10}$). At the same time, granules with 8 to 20 mm length dominated in the fuel under study, their mass fraction was 75.19% (Fig. 2), and the chips content was about 6%. The wood pellets mechanical strength (DU = 98.43%) was slightly lower than the sample taken from the manufacturer’s wood pellet preparation shop No.2 (DU = 98.45%).

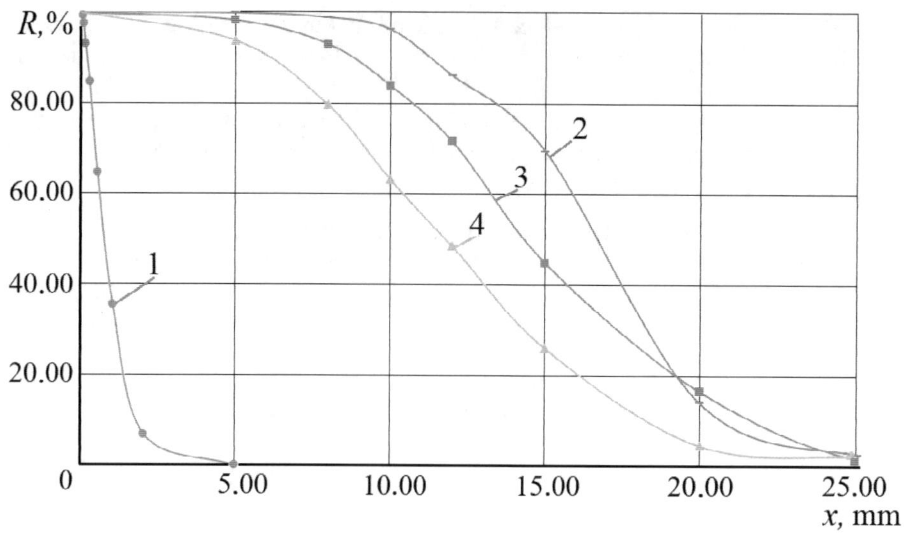

Figure 2. Wood material particle size distribution before the pellet press (1) and pellets shipped by the manufacturer (2), and before the burners of boilers with a rated power of 0.74 MW (3) and 1.5 MW (4).

The study of the particle size distribution, strength and abrasion of wood pellets taken from the auger feeder in front of the boiler with 0.74 MW nominal power showed that the pellets have a less homogeneous and finer composition ($n = 3.105$; $b = 7.863 \cdot 10^{-14}$) than the pellets in Shop No.2. In this case, the studied fuel is dominated by granules with a length of 10 to 25 mm, their mass fraction was 82.51% (Fig. 2), and the crumb content was about 1.6%. The mechanical strength of wood pellets (DU = 98.40%) was slightly lower than in sample taken from the manufacturer in Shop No.2.

The obtained results allow us to conclude that the pellets strength characteristics do not significantly decrease in the systems of wood pellets storage and fuel supply in boiler units equipped with Hekotek boilers rated for 0.74 and 1.5 MW capacity. Moreover, they fully comply with Russian and European standards set for granules of all classes [9]. Earlier, a similar conclusion was made for hot water boilers KBT 4000T, installed in the settlement of "Katunino". However, it is necessary to note the increased content of fines in the samples, especially in the granular fuel for boilers with 1.5 MW nominal power.

The issue of various impurities in wood pellets possible presence, such as bark and sand. As per thermal analyses results, the ash content in dry weight is 0.55% in Shop No.2 granules; thus, the present parameter meets the requirements of Russian and European standards for the highest quality class A1 pellets [9]. This result would be impossible in case of sand in the wood material. The analyses of the Shop No.2 granules structure and color indicates the minimum bark presence in wood material, which is confirmed by thermal analysis results. These facts make it possible to exclude the option of various impurities, such as bark and sand, presence in wood granules and the associated slagging.

Let us mention the issue of boilers operating with significant deviations from the optimal modes. Boiler operators log books review led us to the following conclusions: rarefaction in the boiler furnaces had values close to optimal; flue gas temperatures had unacceptably high values (235–248 °C), while the boilers power was less than 0.61 of the nominal rates. The temperatures were about 120 °C higher than in modes with clean heating surfaces. It should be kept in mind that the temperature of flue gases in front of the smoke exhauster should not exceed 220–250 °C according to the condition of its reliable operation. Thus, boilers were operated under the conditions that are critical for smoke exhausters for a long period.

In response to the question when should the boiler unit be cleaned? It is impossible to give a definite answer, because a lot depends on the conditions for power generation and on the availability of the standby units to swich to if required in winter. Therefore, the time for cleaning the heating surfaces is selected depending on the circumstances. The main rule is: if the flue gases temperature at the same boiler load increased by more than 40–50 °C compared to the one recorded immediately after the soot and ash particles removal, it is time to repeat the cleaning procedure [1].

Based on the foregoing, the operation of hot water boilers with extremely contaminated heating surfaces causes a decrease in their gross efficiency by almost 10%, which accordingly requires an increase in fuel consumption of ~ 10% to ensure the required heating capacity. The increased fuel consumption increases the level of the solid phase in the boiler burners to a value at which the nozzle openings for the secondary air supply are blocked. In boilers with 1.5 MW nominal power, the secondary air nozzles inner diameter is 20 mm, they are 115 mm spaced and their total number is n = 13 + 10 = 23 pcs. In the slag with vitreous formations, removed from the side wall, a ~ 20 mm diameter hole can be seen (Fig. 3), is was obviously "pierced" in the slag melt by the secondary air flow. This vitreous formation in the solidified state is stronger than steel. When the secondary air nozzles are blocked, the boiler automation increases the primary air flow rate to ensure the specified oxygen concentration, that increases the temperature in the burner to a level of fast slagging.

Figure 3. Slag build-up with vitreous formations and a hole.

These circumstances let us conclude that the boilers at the inspected power facilities were operated with the significant deviations from the optimal modes.

Wood pellets from Shop No.2 are delivered to boiler houses by trucks and are loaded into the fuel bunkers. The loading augers move the pellets to elevators and then feed to silos. Visual inspection of the bunkers discovered that a large number of partially and completely destroyed granules accumulated near their walls. It is highly probable that all these fines and dust fall into the silo when the new pellets batch arrive.

Visual examination of the status and color of ash and slag agglomerates removed from the burners showed that their segments differ in color significantly. Based on this finding, the samples were taken from areas with: - black (sample code 09-P), - gray-white (10-P), - violet (12-P) color.

The elements analysis showed a very low combustible substances presence in the samples (Table 2).

The sample taken from the black zone had an increased content of copper oxide (CuO = 0.031%). The sample taken from the violet zone had an increased content of copper and chromium oxides (CuO = 0.028%; Cr_2O_3 = 0.188%). The sample taken from the gray-white zone had an increased content of zinc oxide.

Slag sampling from the zone with vitreous formations (sample code 11-P) for elements composition analyses was very hard, since its harder than steel. The very high strength of this slag zone can be associated with an increased content of vanadium and molybdenum oxides.

Summarizing the above, it should be noted that slagging physical and chemical processes are extremely complex. This is due to the multiphase structure of deposits, consisting of a liquid phase and a number of crystalline phases. Slag deposits differ in composition even layer by layer, which was confirmed by the performed experiments (Table 2). In most cases, the deposits are the result of adhesion of the glassy and partially crystalline phase of ash when it interacts with the grates or lining surface.

Table 2. Element's composition of focal residues in dry weight, %.

	09-P	10-P	11-P	12-P
SrO	0.297	0.120	0.187	0.157
ZnO	0.009	0.136	0.005	0.002
CuO	0.031	0.019	0.012	0.028
NiO	0.018	0.029	0.015	0.019
Fe_2O_3	15.00	13.70	12.70	12.90
MnO	4.49	1.45	2.06	2.08
Cr_2O_3	0.034	0.086	0.124	0.188
CaO	38.00	14.10	24.70	18.20
K_2O	5.31	5.32	4.29	13.80
SO_3	0.733	0.568	0.160	4.41
P_2O_5	2.00	0.868	1.16	0.960
SiO_2	27.20	58.80	49.30	42.50
Al_2O_3	1.49	1.83	1.56	1.49
MgO	3.91	1.39	2.28	1.77
Na_2O	0.229	0.571	0.248	0.441
V_2O_5	-	0.002	0.011	-
TiO_2	0.072	0.086	0.068	0.102
BaO	0.515	0.173	0.301	0.256
MoO_3	-	-	0.042	-
Rb_2O	0.011	0.021	0.014	0.060
N	0.091	0.267	0.220	0.175
C	0.309	0.423	0.529	0.396

In some cases, the formation of the very first layers is affected by the sublimation and subsequent condensation of vapors of low-melting components of the mineral part of the fuel, which, first of all, include oxides of alkali (K_2O, Na_2O) and alkaline earth metals (MgO, CaO). The studies performed have shown the presence of a significant amount of these oxides. In addition, the formation of the very first layers of deposits can also simply cause aggregation of particles due to the surface energy of highly dispersed fractions.

Due to the primary layer stickiness, a loose secondary layer of ash particles grows on its surface. Due to ash low thermal conductivity, the outer temperature of the secondary layer rapidly increases with an increase in its thickness and can exceed the temperature of the liquid phase formation start corresponding to the temperature of the onset of deformation of the ash sample (t_A). The deposits melting leads to an increase in their density and strength, while the surface stickiness increases and the growth of the layer thickness is accelerated. In some cases, the secondary layer hardening is also possible at temperatures below t_A due to processes leading to particle sintering. These processes are mass transfer between particles by the mechanism of molecular diffusion and topochemical reactions. The temperature of the onset of ash deformation characterizes the boundary beyond which intensive slagging of the boiler elements is possible.

Thus, boiler slagging is a progressive accumulation of focal residues in the furnace, in the burner and in the gas ducts, accompanied by the hardening of these deposits due to their melting and sintering. Slagging speed depends on the one hand, on the boiler unit design and operating mode and, on the other hand, on the composition and specific properties of the mineral part of the combusted fuel.

Focal residues, being multicomponent systems, do not have one strictly defined melting temperature. Their transition from solid to liquid state and back occurs in a certain temperature range [5, 6]. The lower boundary of this interval - the solidus temperature corresponds to the appearance of the first drops of liquid melt; a further increase in temperature is accompanied by a decrease in the amount of the solid phase and an increase in the liquid phase. At the liquidus temperature, the last crystals of the solid phase disappear.

Almost all the components of ash and slag are quite refractory. For example, the melting temperature of SiO_2 is 1728 °C, Al_2O_3 is 2050 °C, Fe_2O_3 is 1652 °C, CaO is 2585 °C, MgO is 2800 °C. However, for two- and multicomponent systems, due to the formation of eutectic alloys in them, the emergence of a liquid phase is observed at temperatures significantly below the melting temperatures of pure components

5. Conclusions

Based on the results of the laboratory studies and the analysis of the operating modes of hot water boilers, the following conclusions were made:

1. The main cause of hard ash and slag deposits is the operation of boilers with significant deviations from their optimal modes.
2. The design of some hot water boilers burners does not provide continuous stable withdrawal and tedding of the focal residues accumulated in case the boilers operating modes deviate from the optimal ones; that contributes to fast accumulation of focal residues in the burner and on the combustion chamber lining, accompanied by deposits harderning due to their melting and sintering. These problems were identified for 0.5 to 1.5 MW nominal power Hekotek hot water boilers fueled with high quality wood pellets.
3. All cases of dust cloud occurred during granules loading into the incoming fuel bunkers at the boiler facilities should be recorded and investigated. Regular sampling and monitoring of the incoming biofuel quality should be carried out.
4. It is necessary to maintain the boilers optimal parameters in all operating ranges of load variation in accordance with the Operating Charts. Do not allow to fill the burner and the combustion chamber with wood pellets to the secondary air nozzles level.
5. To provide hot water boilers better and regular cleaning. Temperature of the flue gases leaving the boilers is the most important criteria for the necessity to clean the boilers. To include into the boiler maintenance checklist the steps for cleaning the fuel storage systems and fuel supply conveyors from wood dust, sawdust and "fines" for the purpose: to reduce their share in the burned fuel; to reduce sintering; to reduce their deposition on the lining and on heat exchange surfaces; to exclude their incomplete burning in the boilers increased heat output.
6. When designing boiler facilities for Russian Federation Arctic zone, preference should be given to boilers equipped with combustion chambers, so that they do not depend on the possible slag accumulation because of granular fuel bad quality.

References

[1] Lyubov V K and Lyubov S V 2017 Increasing the efficiency of energy use of biofuels (Arkhangelsk: NArFU) p 533
[2] Berdin V K 2020 *Eur J Polit Econ* **15** pp 106-35
[3] Porfiriev B N and Roginko S A 2016 *Her Russ Acad Sci* **1** pp 433-40
[4] Lyubov V K, Popova E I and Popov A N 2016 *Conference Proceedings 2nd International Conference on Atmospheric Dust* pp 72-77

[5] Beloselsky B S and Solyakov V K 1980 *Energy fuel* (Moscow: Energy) p 168
[6] Zalkind I Ya, Vdovchenko V S and Dick E P *Ash and slag in boiler furnaces* (Moscow: Energoatomizdat) p 80
[7] Trembovlya V I, Finger E D and Avdeyeva A A 1991 *Heating Tests of Boilers* (Moscow: Energoatomizdat) p 416
[8] Alekhnovich A N, Bogomolov V V and Artemiev N V 2001 *Teploenergetika* **2** pp 26-33
[9] Lyubov V K, Malygin P V, Popov A N and Popova E I 2015 *Testing of solid biofuels* (Arkhangelsk: NArFU) p 84

Calculation features of compressed-bent build-up timber columns with nonlinear-deformable shear bracings

E V Popov[1], A V Karelsky[1], V V Sopilov[1], B V Labudin[1] and V V Cherednichenko[1]

[1]Northern (Arctic) Federal University named after M.V. Lomonosov, Department of Engineering Structures, Architecture and Graphics, 163002, Arkhangelsk, Russian Federation

Abstract. Object of research is build-up compressed–bent and eccentrically compressed columns on yielding nonlinear – deformable shear bracings. Purpose of the research is development of a numerical method for calculation of columns, allowing to take in account the influence of deflection of elastic axis of bar on the increment of the bending moment from the action of longitudinal compressive force and the nonlinear dependence between the forces and deformations in the shear bracings. Problem-solving method consists in dividing the column into separate sections, a system of equations is compiled from the condition of equality of the increment of concentrated shears. The loading process is divided into a set number of stages, at each forces in the shear bracings, the stresses in the branches, and the buckling function of the elastic axis of the element are determined. The obtained values of forces in the shear bracings and buckling are used to specify stiffness of the bracings and component of the bending moment arising due to eccentric application of the longitudinal compressive force when longitudinal axis of the element is deflecting. To obtain the resulting values, the obtained forces, deflections and stresses in the branches at each calculation stage are summed up.

1. Introduction

Structural timber and timber–composite structures are used as raw material for production of various types of building structures due to the fact, that such structures have high architectural merit. They are reliable, strong, durable, and at the same time such structures are lightweight. Structural timber is a very economical material, which resistant to aggressive environments. In many countries of the world there is a huge renewable raw material base to produce such structures. The use of timber in the construction of public, industrial, agricultural, multistory residential and store frame buildings is becoming more and more popular. Generally, the major load–bearing elements of such buildings are timber or timber–composite columns.

The basis of modern timber frame buildings is made up of two–hinged and three–hinged frameworks, the basic vertical load–bearing elements of which are timber build–up and lattice columns (Figure 1). Several new design efforts [1] have made it possible to design such columns with a rigid joint in the support section.

IOP Conf. Series: Materials Science and Engineering 1211 (2022) 012007 doi:10.1088/1757-899X/1211/1/012007

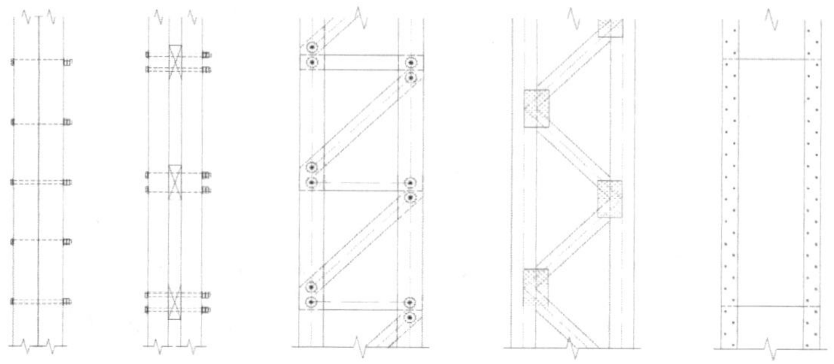

Figure 1. Build–up timber columns: a – with bolt connections; b – the same with gaskets; c – lattice column with combined bolt and claw washer connections; d – the same with metal–toothed plate (MTP) connections; e – I-section column with oriented strand board (OSB) wall

Despite extensive research in the field of calculations, design and testing of build–up timber and timber–composite columns, they are applied only to structures with rigid shear bracings between the composition layers. There are no studies of the behavior of such structures in the presence of compliance of discrete bracings.

When calculating such structures, it is important to consider the curvature of the elastic axis of the column under the action of transverse forces. For build-up columns with yielding bracings this issue deserves special attention because the bending stiffness of composite elements is much lower than for columns with solid cross-section. This fact significantly affects the increase of the bending moment component under the action of a longitudinal compressive force. The method for calculating build-up columns on yielding bracings, given in [2], assumes the use of a linear stiffness coefficient ku. This coefficient does not allow taking into account the real nature of the deformation of connections and thereby the redistribution of forces due to a decrease in the stiffness of more loaded connections, and a decrease in the bending stiffness of the entire structure at loading it.

The object of our research was compressed–bent build–up timber columns. The subject of the research was non–linear dependence «load-shear» of bracings and the deformability of such structures. The purpose of the article was to design a step method for calculation, which considers the increase of bending moment component because of eccentric compressive force, and non–linear deformability of shear bracings. The main objective of our research was to calculate a two–layer build–up timber column according to the developed mathematical model, and to compare the exact and approximate calculations of such a structure with the assumption of a different number of iterations and a different approach for determining the elastic modulus of bracings.

2. Materials and methods

There is a nonlinear dependence between the force value and the deformation of the connection for most types of shear bracings: dowels [3], bolts [4], MTP [5], claw connectors [6], screws [7], clamps [8], etc. Thereby, it is necessary to consider the change in stiffness depending on the forces for build-up columns with nonlinearly deformable yielding connections. The stiffness coefficient of each bracing should be considered as a function [9]:

$$c = c\left(T_c\right) \tag{1}$$

where c is shear stiffness coefficient of the single bracing; T_c is the force per bracing.

The scheme of build–up column is shown in Figure 2. Forces T_c are arising in the shear bonds under the action of a bending moment, the nature of the distribution of which depends on the form of the end restraint and the load case. When the load is transferred to the column centrally in the initial undeformed state, the longitudinal force is distributed between the branches in proportion to their longitudinal stiffness EF, and the forces in the shear bracings is equal to T_c=0. Under the action of a transverse load $q(z)$ (where z is the coordinate, measured along the height of the bar), the column axis is beginning to deflect from the vertical, a bending moment $M_{0,q}$ from the transverse load $q(z)$ application and a bending moment $M_{0,N}$ appears because of eccentric longitudinal force $\sum N$ application, $M_{0,N}=e(z)\cdot\sum N$ (where $e(z)$ is the eccentricity function). Here and further in the text, the index "0" means that the force factor is determined without taking account of the shear forces in the discrete bonds.

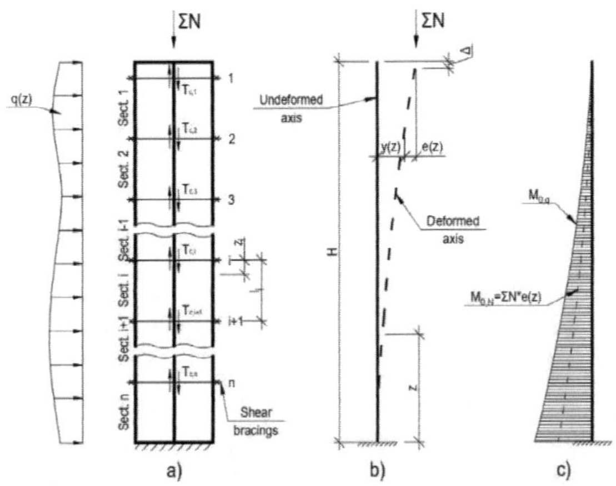

Figure 2. Compressed–bent composite element with yielding discrete shear bracings: *a* is numbering of bracings and sections; *b* is a scheme of the deformation of the column under the combined action of a compressive force and transverse load; *c* is an epure of bending moments $M_{0,q}$ an $M_{0,N}$

Let us consider *i*–th section. The increment of the concentrated shear along the length of the selected section is equal to the difference between the shears of the *i*–th and *i*–1–th bracings:

$$\Gamma_i - \Gamma_{i-1} = \frac{T_{c,i}}{c_i} - \frac{T_{c,i-1}}{c_{i-1}} = \gamma \sum_{k=i}^{n} T_k + \int_0^{l_i} \Delta_i(z_i)dz_i \tag{2}$$

where c_i is the stiffness coefficient of the *i*–th bracing; z_i is coordinate, which measured along the length of the *i*–th section; γ, Δ_i are parameters, which determined by the formulas:

$$\gamma = \frac{1}{E_1F_1} + \frac{1}{E_2F_2} + \frac{w^2}{\Sigma EI}; \qquad \Delta_i(z_i) = \frac{N_2}{E_2F_2} - \frac{N_1}{E_1F_1} - \frac{M_{0,i}(z_i)\cdot w}{\Sigma EI} \tag{3}$$

где F_1, F_2, E_1; E_2 are cross sections and elastic moduli of branches material of build-up column; N_1; N_2 are longitudinal forces in branches; w is the distance between gravity centers of cross section of branches; $\sum EI$ is the sum of the stiffness of the branches: $\Sigma EI=E_1I_1+E_2I_2$; $M_{0,i}(z_i)$ is the function of bending moment distribution within the *i*–th section: $M_{0,i}= M_{0,I,q}+ M_{0,i,N.}$

Equations (2) allows to get for each section a system of equations for determining the shear forces in bracings. The end point should be taken as the section in which the concentrated shear $\Gamma=0$. For example, for a rigidly supported console, such section is the support section; for a column pin–supported at the ends without obstacles to shear is the point at which the epure Q_0 changes sign. In this case, for asymmetric loading schemes, systems of equations should be compiled separately for each column parts below and above the section without concentrated shear.

$$\begin{cases} \dfrac{T_{c,2}}{c_2} - \dfrac{T_{c,1}}{c_1} = \gamma \cdot T_1 \cdot l_1 + \int\limits_0^{l_1} \Delta_1(z_1)dz_1 \\ \dfrac{T_{c,3}}{c_3} - \dfrac{T_{c,2}}{c_2} = \gamma \cdot \sum\limits_{k=1}^{2} T_k \cdot l_2 + \int\limits_0^{l_2} \Delta_2(z_2)dz_2 \\ \dots \\ \dfrac{T_{c,i+1}}{c_{i+1}} - \dfrac{T_{c,i}}{c_i} = \gamma \cdot \sum\limits_{k=1}^{i} T_k \cdot l_i + \int\limits_0^{l_i} \Delta_i(z_i)dz_i \\ \dots \\ \dfrac{T_{c,n}}{c_n} - \dfrac{T_{c,n-1}}{c_{n-1}} = \gamma \cdot \sum\limits_{k=1}^{n-1} T_k \cdot l_{n-1} + \int\limits_0^{l_{n-1}} \Delta_{n-1}(z_{n-1})dz_{n-1} \\ -\dfrac{T_{c,n}}{c_n} = \gamma \cdot \sum\limits_{k=1}^{n} T_k \cdot l_n + \int\limits_0^{l_n} \Delta_n(z_n)dz_n \end{cases} \tag{4}$$

The system of equations (4) can be represented in matrix form:

$$X = A^{-1} \cdot B \tag{5}$$

where X is the matrix of unknown shear forces $T_{c,i}$; A is the matrix of coefficients of unknown shear forces $T_{c,i}$ (formula (6)); B is the matrix of free terms (integrals in the right side of equations (4)).

$$A = \begin{bmatrix} -\dfrac{1}{c_1} - \gamma l_1 & \dfrac{1}{c_2} & 0 & \dots & 0 & \dots & 0 \\ -\gamma l_2 & \dfrac{1}{c_2} - \gamma l_2 & \dfrac{1}{c_3} & \dots & 0 & \dots & 0 \\ \dots & \dots & \dots & \dots & \dots & \dots & \dots \\ -\gamma l_i & -\gamma l_i & -\gamma l_i & \dots & -\dfrac{1}{c_i} - \gamma l_i & \dots & 0 \\ \dots & \dots & \dots & \dots & \dots & \dots & \dots \\ -\gamma l_n & -\gamma l_n & -\gamma l_n & \dots & -\gamma l_n & \dots & -\dfrac{1}{c_n} - \gamma l_n \end{bmatrix}; \; B = \begin{bmatrix} \int\limits_0^{l_1} \Delta_1(z_1)dz_1 \\ \int\limits_0^{l_2} \Delta_2(z_2)dz_2 \\ \dots \\ \int\limits_0^{l_i} \Delta_i(z_i)dz_i \\ \dots \\ \int\limits_0^{l_n} \Delta_n(z_n)dz_n \end{bmatrix}; \; X = \begin{bmatrix} T_{c,1} \\ T_{c,2} \\ \dots \\ T_{c,i} \\ \dots \\ T_{c,n} \end{bmatrix} \tag{6}$$

The solution of the system of equations gives the values of the shear forces in bracing, and the considered scheme becomes statically definable. The stresses in the branches are determined according to the common structural mechanics rules.

The complexity of this solution in an analytical form for compressed–bent columns with nonlinear-deformable shear bracings is that the bending moment in the sections of the column can be calculated only when the horizontal displacements are known, at the same time, the latter cannot be determined without values of the bending moments and shear bracings forces, the stiffness of which, in turn, depends on the shear forces acting in them. A numerical algorithm for solving such a problem, which consists in the following, is proposed in this paper:

– the transverse load $q(z)$ is applied in steps of δq, and the longitudinal compressive force at the initial stage of the calculation has a full value due to the fact, that longitudinal force in the columns, as a rule, is created due to constant, useful and snow loads, which are long–acting in nature, in contrast to the transverse load q, which usually arises due to the wind pressure;

– at each stage of the calculation, the forces in the shear bracings, the stresses in the branches and the displacements of the bar are determined. The displacements values is used to refine the component of the bending moment $M_{0,N}$ for the next stage. The stiffness coefficients of the bracings c_i are refined for the next step, depending on the magnitude of the acting shear forces, which are unevenly distributed between the bracings;

– stresses in the branches and forces in shear bracings are summed up at each stage of the calculation, based on the resulting values of which conclusions about ensuring the strength and rigidity of the structure are formulated.

The common algorithm for axis displacements of a build–up structure determining, considering the nonlinear compliance of shear bracings, is described in [10].

The equation of the bending line of the section between the top of the column and bracing 1 (section 0) is as follows:

$$y_0(z_0) = \frac{1}{\Sigma EI} \iint \left[M_{0,q}^{j}(z_0) + \sum N \cdot e^{j-1}(z_0) \right] dz_0 dz_0, \tag{7}$$

For other sections (i=1, 2…n):

$$y_i(z_i) = \frac{1}{\Sigma EI} \iint \left(M_{0,q}^{j}(z_i) + \sum N \cdot e^{j-1}(z_i) + \sum_{k=1}^{i} T_k \cdot w \right) dz_i dz_i, \tag{8}$$

where $e^{j-1}(z)$ is the eccentricity function of the application of the longitudinal force, calculated at the j–1 stage of the calculation.

The solutions to equations (7) and (8) will be presented in the form:

$$y_i(z_i) = \frac{1}{\Sigma EI} \left(\Phi_i(z_i) + C_i z_i + D_i \right), \quad \left(i = 0,\ 1 \ldots n \right) \tag{9}$$

where $\Phi_i(z_i)$ are functions that are the general solution of indefinite integrals of expressions (7) and (8); C_i, D_i are arbitrary constants.

To determine arbitrary constants, a system of equations is drawn up that connects deflections and angles of rotation at the boundaries of the sections (continuity conditions). For a column rigidly clamped in the support section, considering the equality of the deflection and the angle of rotation in it to zero, the system of equations can be written in the form:

$$\begin{cases} \Phi_0(l_0) + C_0 \cdot l_0 + D_0 = \Phi_1(0) + D_1 \\ \Phi_0^{'}(l_0) + C_0 = \Phi_1^{'}(0) + C_1 \\ \Phi_1(l_1) + C_1 \cdot l_1 + D_1 = \Phi_2(0) + D_2 \\ \Phi_1^{'}(l_1) + C_1 = \Phi_2^{'}(0) + C_2 \\ \qquad \ldots \\ \Phi_{n-1}(l_{n-1}) + C_{n-1} \cdot l_{n-1} + D_{n-1} = \Phi_n(0) + D_n \\ \Phi_{n-1}^{'}(l_{n-1}) + C_{n-1} = \Phi_n^{'}(0) + C_n \\ \Phi_n(l_n) + C_n \cdot l_n + D_n = 0 \\ \Phi_n^{'}(l_n) + C_n = 0 \end{cases} \tag{10}$$

The system of equations can be represented in matrix form (5):

$$A=\begin{bmatrix} l_0 & 1 & 0 & 1 & 0 & 0 & & 0 & 0 & 0 & 0 \\ 1 & 0 & 1 & 0 & 0 & 0 & & 0 & 0 & 0 & 0 \\ 0 & 0 & l_1 & 1 & 0 & 1 & & 0 & 0 & 0 & 0 \\ 0 & 0 & 1 & 0 & 1 & 0 & \ldots & 0 & 0 & 0 & 0 \\ & & & & \ldots & & & \ldots & & & \\ 0 & 0 & 0 & 0 & 0 & 0 & \ldots & l_{n-1} & 1 & 0 & 1 \\ 0 & 0 & 0 & 0 & 0 & 0 & & 1 & 0 & 1 & 0 \\ 0 & 0 & 0 & 0 & 0 & 0 & & 0 & 0 & l_n & 1 \\ 0 & 0 & 0 & 0 & 0 & 0 & & 0 & 0 & 1 & 0 \end{bmatrix};\ B=\begin{bmatrix} \Phi_1(0)-\Phi_0(l_0) \\ \Phi_1'(0)-\Phi_0'(l_0) \\ \Phi_2(0)-\Phi_1(l_1) \\ \Phi_2'(0)-\Phi_1'(l_1) \\ \ldots \\ \Phi_n(0)-\ldots \\ \Phi_n'(0)-\ldots \\ -\Phi_n(0) \\ -\Phi_n'(l_n) \end{bmatrix};\ X=\begin{bmatrix} C_0 \\ C_1 \\ D_0 \\ D_1 \\ \ldots \\ C_{n-1} \\ D_{n-1} \\ C_n \\ D_n \end{bmatrix} \qquad (11)$$

The required horizontal displacement function is found by approximating the obtained values at the characteristic points.

As an example, a build-up timber column of the cross frame of a one–storey building with a timber frame (Figure 3, a) made of pine wood of C22 strength class with a modulus of elasticity E=6.7 GPa is considered. The branches of the build-up column are connected by steel bolted joints and toothed connectors. The deformation of the joints occurs according to a nonlinear behavior, the stiffness parameter of the joints is determined by the dependence $c=c(T_c)$ (Fig. 3, b). The deformation of structural timber is assumed to be linearly elastic. The parameters, which are taken as initial data: column height H=3.2 m, cross–sectional dimensions of branches ($b\times h$) 200×150 mm, spacing of shear bracings 0.5 m. The column is loaded with a longitudinal force $\sum N$=200 kN, transmitted to the column centrally through a special cap. A uniformly distributed transverse wind load with intensity q=5 kN/m acts on the column. It is required to determine the shear forces in bracings, displacements of axis and maximum edge stresses in the branches of a build-up column.

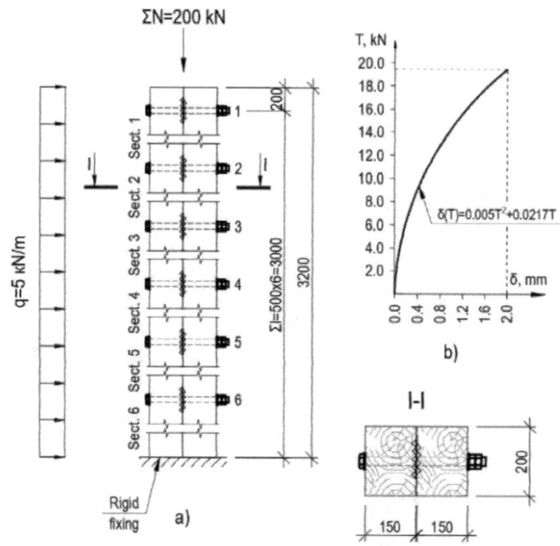

Figure 3. The calculation model of a build–up column with yielding bracings: *a* – column design; *b* – diagram «load–deformation» (T – δ) dependence for a single bolted bracing under the action of longitudinal shear

The column is separated into 7 sections of length l_j. Shown below the functional dependences of bending moments distribution along the length of the sections $M_{0,i}(z_i)$ at the j–th stage of the calculation:

– Section 0:

$$M_0^j(z_0) = \frac{qz_1^2}{2} + \Sigma N \cdot e^{j-1}(z_0) \tag{12}$$

– Section 1:

$$M_1^j(z_1) = q\left((H - \Sigma l)\left(\frac{H - \Sigma l}{2} + z_1\right) + \frac{z_1^2}{2}\right) + \Sigma N \cdot e^{j-1}(z_1) \tag{13}$$

– Section i (i=2…6):

$$M_i^j(z_i) = q\left((H - \Sigma l)\left(\frac{H - \Sigma l}{2} + \sum_{k=1}^{i-1} z_k + z_i\right) + \sum_{k=1}^{i-1} z_k\left(\frac{\sum_{k=1}^{i-1} z_k}{2} + z_i\right) + \frac{z_i^2}{2}\right) + \Sigma N \cdot e^{j-1}(z_i); \tag{14}$$

To determine the forces in the shear bracings at each stage of the calculation, a system of equations (4) is drawn up, then - a system of equations for determining arbitrary constants C_0, D_0, … C_6, D_6 according to the expression (9). Horizontal displacements are calculated at distinguished points along the column height: at the top level and at each bracing. Horizontal displacements are determined by formula (9) at points corresponding to the beginning and end of the sections. The approximating buckling functions are taken in the form of 3-degree polynomials, allowing to obtain a high degree of accuracy of the approximation (R≥0.995):

$$\overline{y_i}(z_i) = a_j z_i^3 + b_j z_i^3 + c_j z_i; \quad (i = 0,\ 1\ \dots\ 6) \tag{15}$$

where a_j, b_j, c_j are polynomial coefficients calculated by the j-th stage of the calculation.

The eccentricity function e^j, at the j-th stage of calculation, corresponding to the column deformation scheme shown in Fig. 2b will be of the form:

$$e_i^j(z_i) = \overline{y_0}(0) - \overline{y_i}(z_i) \tag{16}$$

The stiffness coefficients of the bracings at the j–th stage of calculation are determined by the forces, which obtained at the j–1–th stage of the calculation and are refined by the formula:

$$c_i^j(T_i^{j-1}) = \frac{1}{\mathrm{d}\delta_i\left(\sum_{k=1}^{j-1} T_{i,k}\right) / \mathrm{d}T}. \tag{17}$$

where $\sum_{j=1}^{k-1} T_{i,k}$ is the total value of shear forces in the i–th bracing, obtained at the previous j–th stages of the calculation; $\delta_i(T)$ is deformation of the i–th bracing at a assumed load, determined by the approximating curve of the graph of the deformation of the connection in Fig. 3.

The edge normal stresses σ_x in the relevant cross–section are determined by the formula:

$$\sigma_x = \frac{\sum N \cdot F_1}{F_1 + F_2} + \sum_{j=1}^{m} \sigma_{x,M}^j \tag{18}$$

where $\sigma_{x,M}^j$ are edge normal stresses at the j–th stage of the calculation, which obtained without considering the action of the longitudinal compressive force $\sum N$ (considered only when calculating the bending moment), determined for 1st and 2nd branches, respectively, by the formulas:

$$\sigma_{x,M}^j = \pm \frac{F_1 \sum_{k=1}^{i} T_k}{F_1 + F_2} \pm \frac{\left(M_0^j - w\sum_{k=1}^{i} T_k\right) I_1 \frac{h_1}{2}}{I_1 + I_2}; \quad \sigma_{x,M}^j = \pm \frac{F_2 \sum_{k=1}^{i} T_k}{F_1 + F_2} \pm \frac{\left(M_0^j - w\sum_{k=1}^{i} T_k\right) I_2 \frac{h_2}{2}}{I_1 + I_2} \tag{19}$$

where $\sum_{k=1}^{i} T_k^j$ is the total value of the forces in the shear bracings above the considered cross-section at the j–th stage of the calculation; M_0^j is bending moment, obtained in the considered cross-section at the j–th stage of the calculation without taking into account shear forces in the bracings; h_1, h_2, F_1, F_2, I_1, I_2 are cross–section height, cross–sectional area and moments of inertia of the 1st and 2nd branch, respectively.

The calculation is performed with the number of stages: m=1, 2, 5, 10, 15 and 20. For the first stage of loading, the stiffness coefficient of the connection is taken equal to the tangent of the angle of arrival, drawn through the initial point of the graph to the axis of abscissae (δ). At this stage, horizontal displacements are determined only by the action of a lateral load. At each stage of the calculation, the values of the forces in bracings T_i are determined from the solution of system of equations (5) and the horizontal displacements of the top of the column. According to the data obtained, the parameters of the stiffness of the connections c_i and the coefficient α^j, considering the influence of the buckling of the element on the bending moment, are refined for the next stage of the calculation. The values of the shear forces in the bracings to clarify the stiffness coefficients c_i at the next stage of the calculation are taken equal to the sum of the forces at the previous stages of the calculation.

To compare the results, a calculation is made at a constant value of the stiffness coefficient of connection k_u, which, according to [13], is determined by the formula (22) for bolted joints with toothed connectors:

$$k_u = \frac{2}{3} k_{ser}, \tag{20}$$

where k_{ser} is norm stiffness coefficient of connection, defined as a secant modulus at a load equal to 40% of the admissible limit value.

3. Results and discussion

The graphs in Figure 4 shows the values of the shear forces in the bracings for a different number of calculation stages m. The variant of calculation with m=20 is hereinafter accepted as a «reference» one.

In the linear calculation (m=1, the forces in the shear bracings are determined without considering the eccentric application of the longitudinal force, the stiffness coefficient of the bracings is determined by formula (20)) the error is 30...42%. A further increase in the number of iterations gives the following errors: 2...26% at m=2; 1...7% at m=5; 1...3% at m=10; 0.5...1% at m=15. Calculation with a linear value of the stiffness coefficient, calculated by the formula (20), with the number of iterations m=10 gives an error of 3...22%. Thuswise, to obtain reliable values of the shear forces in the bracings, which are necessary for checking the strength of the connections, m=5...10 iterations should be enough.

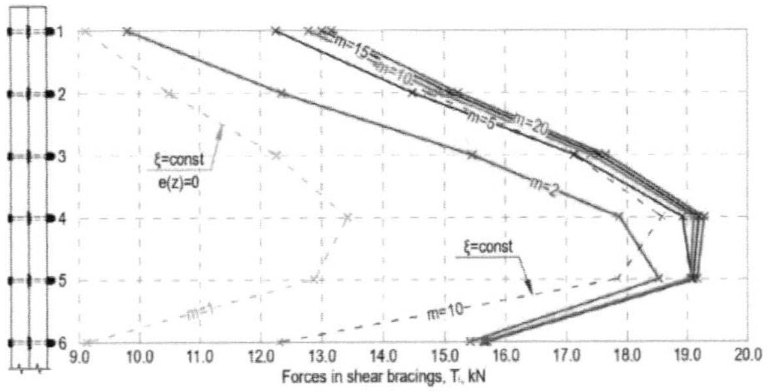

Figure 4. Forces in shear bracings T_i for different numbers of iterations m

The graphs in Figure 5 shows the values of horizontal displacements. In a linear calculation (m=1, the forces in the shear bracings are determined without considering the eccentric application of the longitudinal force, the stiffness coefficient of the bracings is determined by formula (20)), the error is 18%. A further increase in the number of iterations gives the following errors: 26% for m=2; 10% at m=5; 4% at m=10; 1.5% at m=15. Calculation with a linear value of the stiffness coefficient calculated by formula (02), with the number of iterations m=10, gives an error of 14%.

It should be noted that the difference in indicators in favor of a linear calculation at m=1 compared to nonlinear at the number of iterations m=2 is only a special case, since it is due to the underestimated value of the linear stiffness coefficient of the shear bracings, which introduced into the calculation.

The accuracy of the obtained values of horizontal displacements is important not only for the deformability of a structure estimation, but also for an adequate material strength assessment and shear bracings. Deflection of the longitudinal axis of the element leads to a significant increase in the component of the bending moment $M_{0,N}$. This is especially important to consider when calculating build–up columns with yielding bracings, which deformability is much higher than solid elements with the same cross-section one. To obtain reliable results of horizontal displacements, it is recommended to use at least m=10 iterations.

Figure 6 shows the of normal stresses epures in the branches in the cross-section of the rigid fixing of the column. In a linear calculation (m=1, the forces in the shear bracings are determined without considering the eccentric application of the longitudinal force, the stiffness coefficient of the bracings is determined by formula (20)), the error is 11...29%. A further increase in the number of iterations gives the following errors: 8 ... 28% at m=2; 3...9.5% at m=5; 1...4% at m=10; 0.5...1.5% at m=15.

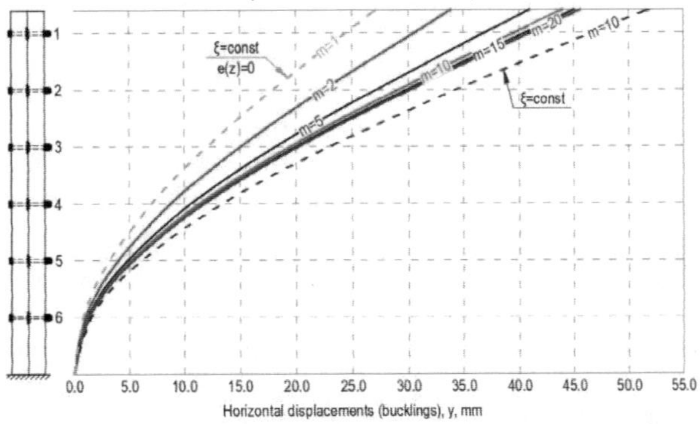

Figure 5. The deflection graphs of the elastic axis of the column $y(z)$ during transversive-longitudinal bending

Calculation with a linear value of the stiffness coefficient, calculated by the formula (20), with the number of iterations m=10 gives an error of 6...26%. The largest dispersion of values is typical for the inner edge of the second branch. The leap of stresses is due to the compliance of shear bracings. With

 doi:10.1088/1757-899X/1211/1/012007

a high stiffness of the shear bracings and a small spacing ($c_i \to \infty$), the difference in the values of normal stresses on the internal edges of the branches tends to zero. In this case, the strength condition of the structure can be both the values of tensile stresses from the external side of first branch and maximum compressive stresses from the external side of second branch. This is due to the lower tensile strength of timber than compressive strength. To obtain adequate values of normal stresses, it is recommended to use the iteration value m=10.

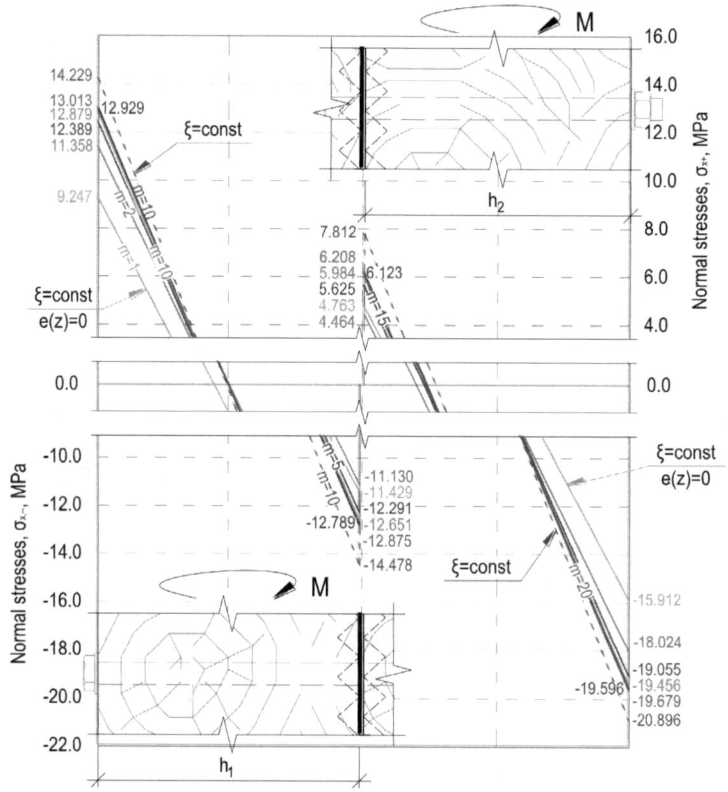

Figure 6. Normal stress distribution diagrams σ_x in the branches of the built-up column

4. Conclusions

1 A numerical methods for compressed–bent and eccentrically–compressed build–up columns with yielding shear bracings calculating is presented. It allows to take the influence of the deflection of the elastic axis of the bar on the increment of the bending moment from the longitudinal compressive force operation and the nonlinear dependence between forces and deformations in the shear bracings. The calculation is performed using a n-step method (the loading process is separated into a adjusted number of stages *m*). Formulas for calculating the shear forces in the bracings, horizontal displacements of the elastic axis and edge normal stresses in branches are obtained.

2 The calculation of a cantilever two–branch compressed–bent timber column with a different number of iterations, as well as in a linear statement. The use of the linear coefficient of stiffness of the connections k_u gives an error in the forces of the shear bracings determining up to 42%, horizontal displacements – up to 26%, stresses in the branches – 11...29%.

3 For a proper evaluation of the stress-strain state of such structures, the number of iterations should be taken m=10. A further increase in the number of iterations gives insignificant divergence in the required values.

5. References

[1] Rimshin V, Labudin B, Morozov V, Orlov A, Kazarian A and Kazaryan V 2019 Calculation of Shear Stability of Conjugation of the Main Pillars with the Foundation in Wooden Frame Buildings Advances in Intelligent Systems and Computing. 983 9 doi:10.1007/978-3-030-19868-8_85

[2] CEN EN 1995–1–1:2004/A2–2014 Eurocode 5: Design of timber structures – Part 1–1: General – Common rules and rules for buildings. URL: https://docs.cntd.ru/document/552380618 (date of application 05.09.2021)

[3] Šmak M and Straka B 2014 Development of new types of timber structures based on theoretical analysis and their real behaviour *Wood Research.* 59(3) 11

[4] Lokaj A and Klajmonová K 2014 Round timber bolted joints exposed to static and dynamic loading *Wood Research.* 59(3) 9

[5] Karelskiy A V, Zhuravleva T P and Labudin B V 2015 Load-to-failure bending test of wood composite beams connected by gang nail *Magazine of Civil Engineering.* 54(02) 8 doi:10.5862/MCE.54.9

[6] Chernykh A, Danilov E and Koval P 2020 Stiffness Analysis of Connections of LVL Structures with Claw Washers Lesnoy Zhurnal (*Forestry Journal*). 4 10 doi:10.37482/0536-1036-2020-4-157-167

[7] Baszeń M 2017 Semi-rigid Behavior of Joints in Wood Light-frame Structures Procedia Engineering. 172 7 doi:10.1016/j.proeng.2017.02.022

[8] Hassanieh A and Valipour H 2020 Experimental and numerical study of OSB sheathed-LVL stud wall with stapled connections *Construction and Building Materials.* 233 doi:10.1016/j.conbuildmat.2019.117373

[19] Labudin B, Popov E, Vladimirova O, Sopilov V and Bobyleva A 2021 Wood-Composite Structures with Non–Linear Behavior of Semi-Rigid Shear Ties Construction of Unique Buildings and Structures. 97 doi: 10.4123/CUBS.97.2

[10] Popov E V, Sopilov V V, Bardin I and Lyapin D M 2021 Calculation of Vertical Deformations of Composite Bending Wooden Structures with Non–linear Behavior of Shear Bonds Environmental and Construction Engineering: Reality and the Future. 160 7 doi:10.1007/978-3-030-75182-1_15

IOP Conf. Series: Materials Science and Engineering 1211 (2022) 012008 doi:10.1088/1757-899X/1211/1/012008

Methodology for determining the flow rate and temperature of generated hot water when calculating the thermal diagram of a hot water boiler room

V P Gorshenin[1], A N Kachanov[1], VA Chernyshov[1], D A Korenkov[1], V V Maksimov[2]
[1] Oryol State University, 95 Komsomolskaya Street, Orel, Russia
[2] Kazan State Tekhnicheskiy University, 51 Krasnoselskaya Street, Kazan, Russia

Abstract: The consumption and temperature of hot water produced by boiler units are calculated using such a quantity as the consumption of the consumed network water. The formulas represent functional dependencies connecting the obtained values not only with the consumption of heating water, but also with the consumption of hot water supplied for recirculation, for own needs and for the needs of the fuel economy. They take into account the losses of network and hot water. The formulas are obtained as a result of the joint solution of the equation for the total consumption of hot water and the equation for its partial costs.

1. Introduction. At present, one of the main tasks facing enterprises of the housing and communal complex , is energy saving - increasing the efficiency of the use of fuel and energy resources. The goals of calculating the heating scheme of a boiler house [1]: determination of its total heat load, including external loads and heat consumption for auxiliary needs; calculation of heat fluxes and flow rates of heat carriers through all types of auxiliary equipment, which ensures its selection and determination of the diameters of pipelines and fittings; determination of the initial data for further technical and economic calculations. The calculation of the heating scheme of the boiler house is carried out for the typical modes of its operation [1, 2]. The calculation results make it possible to correctly substantiate the number and performance of auxiliary equipment.

The basic thermal diagram of a boiler house with hot water boilers is shown in the figure 1.

The boiler room operates on a closed heating network. The pipelines, depicted in the form of dotted lines, ensure the inclusion of heaters of the original 2 and chemically treated 5 water in the heating circuit of the boiler house according to various options. In the presented thermal diagram, heaters 2 and 5 for heating (hot) water are connected in series.

2. Materials and methods

At the outlet of the boiler units, hot water is divided into four streams: into the heating network, for recirculation, for the auxiliary needs of the boiler house and into the fuel economy:

$$G^{bh}_{HW(prod)} = G^{hn}_{HW(prod)} + G_{HW(prod)} + G^{on}_{HW(prod)} + G^{fu}_{HW(prod)}, \quad (1)$$

where $G^{bh}_{HW(prod)}$ – total consumption of hot water produced by the boiler house; $G^{hn}_{HW(prod)}$ – consumption of hot water generated to meet the household needs of residential buildings (for supply to the heating network); $G_{HW(prod)}$ – consumption of hot water generated for its recirculation from the outlet to the inlet of the boiler units; $G^{on}_{HW(prod)}$ – consumption of hot water generated to meet the boiler

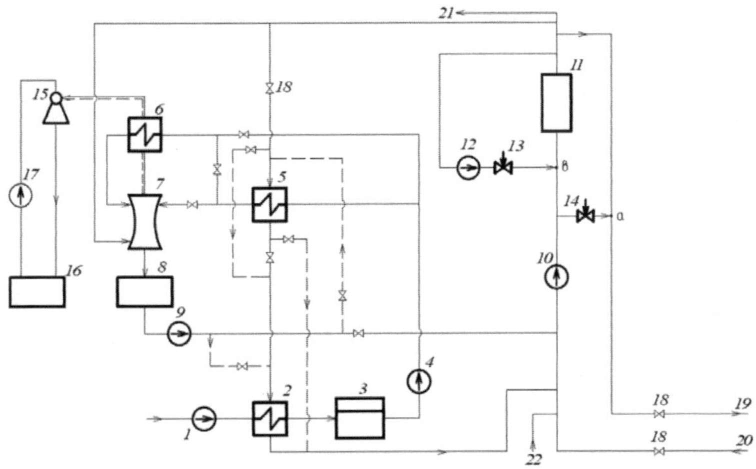

Figure 1. Basic thermal diagram of a heating boiler house with hot water boiler units operating on a closed heating network:

1 - initial (raw) water pump; 2 and 5 - heaters of raw and chemically treated water; 3 - chemical water treatment area; 4 - pump of chemically treated water; 6 - vapor cooler; 7 - vacuum deaerator; 8 - make-up (deaerated) water tank; 9 - make-up water pump; 10 - network water pump; 11 - hot water boiler unit; 12 - recirculation pump; 13 and 14 - control valves; 15 - water jet ejector; 16 - working water tank; 17 - working water pump; 18 - gate valve; 19 and 20 - supply and return heat pipelines of the heating network; 21 and 22 - supply and return heat pipelines of the fuel oil facility.

room's own needs (for supply to deaerator 7 and heat exchangers 2 and 5); $G^{fu}_{HW(prod)}$ – consumption of hot water generated to meet the needs of the fuel economy.

Hot water generated by boiler units, when moving to consumers, is partially lost in the boiler room and in external heat pipelines. In this regard, the flow rate $G^{hn}_{HW(prod)}$ of hot water is represented as follows:

$$G^{hn}_{HW(prod)} = G^{p(s)}_{HW(L)} + G^{n(s)}_{HW(L)} + G^{hn}_{HW(C)} = k^{hn}_{HW} \cdot G^{hn}_{HW(C)} = k^{hn}_{HW} \cdot \alpha_{HW(D)} \cdot G_D, \quad (2)$$

where $G^{p(s)}_{HW(L)}$, $G^{n(s)}_{HW(L)}$ – the consumption of hot water lost in the supply heat pipelines, respectively, of the boiler house and heating network; $G^{hn}_{HW(C)}$ – consumption of hot water consumed to meet the household needs of residential buildings; k^{hn}_{HW} – coefficient taking into account the loss of hot water in the supply heat pipes of the boiler house and heating network; $\alpha_{HW(D)}$ – the share of hot water consumption in the network water consumption; G_D – current consumption of the consumed network water.

The value $\alpha_{HW(D)}$ is determined within the framework of the requirements for the law of temperature change τ_1^* of direct supply water. The change in this temperature during the heating period is carried out in accordance with the temperature schedule of the heating network. In this case, within the framework of the heating season, two ranges of outdoor temperature values are distinguished: $\tau_{ob} \ldots \tau_{obp}$; $\tau_{obp} \ldots \tau_{doa}$ where τ_{ob} – outdoor temperature corresponding to the beginning (end) of the heating period,°C; τ_{obp} – outside air temperature corresponding to the break point of the temperature graph,°C; τ_{doa} – design outside air temperature (for heating design), °C.

To provide at each moment of time the required temperature value τ_1^* of direct heating water, chilled water is added to the hot water. The flow rates of hot water and chilled water mixed with it are determined by jointly solving the equations of material and heat balances written for the mixing points a (see figure). Solving these equations gives:

$$G^{hn}_{HW(mix)} = G^{hn}_{HW(S)} = \alpha_{HW(D)} \cdot G_{D(S)}; \quad (3)$$

$$G^{hn}_{RW(\text{mix})} = G_{D(S)} - G^{hn}_{HW(S)} = \alpha_{RW(D)} \cdot G_{D(S)}, \quad (4)$$

where $G^{hn}_{HW(mix)}$ – consumption of hot water coming from boiler units to the mixing points A and then entering as part of the network water into the supply line of the heating network; $G^{hn}_{RW(\text{mix})}$ – flow rate of the recovered (replenished) return water flowing through the bridges from the boiler units inlet to the outlet pipelines (at the mixing point A); $G_{D(S)}$ – consumption of network water supplied to the supply line of the heating network; $\alpha_{HW(D)}$, $\alpha_{RW(D)}$ – the share of the consumption of hot and recovered return water, respectively, in the flow of heating water ($\alpha_{HW(D)}+\alpha_{RW(D)}=1$).

The quantities $\alpha_{HW(D)}$ and $\alpha_{RW(D)}$ are equal:

$$\alpha_{HW(D)} = \frac{\tau_1^* - \tau_{RW}}{\tau_{2(HW)} - \tau_{RW}}; \quad (5)$$

$$\alpha_{RW(D)} = \frac{\tau_{2(HW)} - \tau_1^*}{\tau_{2(HW)} - \tau_{RW}}, \quad (6)$$

where τ_1^* – temperature of the direct supply water at the entrance to the supply line of the heating network [3]; τ_{RW} – temperature of the recovered hot water return at the outlet of the network pumps, °C; $\tau_{2(HW)}$ – temperature of hot water produced by boiler units, °C.

The τ_{RW} temperature of the recovered return water at the outlet of the network pumps is found as a weighted average:

$$\tau_{RW} = \frac{\alpha_{DW(WVR)} \cdot \tau_{DW} + \alpha^H_{HW(WVR)} \cdot \tau^H_{2(HW)} + \alpha^{CH}_{HW(WVR)} \cdot \tau^{CH}_{2(HW)} + 1 \cdot \tau_2^*}{\alpha_{DW(WVR)} + \alpha^H_{HW(WVR)} + \alpha^{CH}_{HW(WVR)} + 1}, \quad (7)$$

where $\alpha_{DW(WVR)}$, $\alpha^H_{HW(WVR)}$, $\alpha^{CH}_{HW(WVR)}$ – the share of the consumption of deaerated and cooled heating water entering the return line (to the input of the network pumps), respectively, from the deaerator, heaters of chemically treated and source water, fuel facilities, from the consumption of network water returning to the boiler room; τ_{DW} – temperature of deaerated water, at the inlet to the return line, °C; $\tau^H_{2(HW)}$, $\tau^{CH}_{2(HW)}$ – the temperature of the cooled heating water entering the return line from the heaters of chemically treated and initial water, the fuel economy, respectively,°C; τ_2^* τ_2^*– temperature of the heating system return water returning to the boiler room at the inlet of the network pumps [3].

The values $\alpha_{DW(WVR)}$, $\alpha^H_{HW(WVR)}$, $\alpha^{CH}_{HW(WVR)}$ in formula (7) at this stage of calculating the boiler room scheme are not known and therefore are preset. In order to avoid low-temperature corrosion, the design temperature $\tau_{1S(HW)}$ of water at the inlet to steel water-heating boiler units, depending on the type of fuel, should be at least 70 ... 110 °C [4 - 6]. Maintaining the required value of this water temperature is ensured by arranging a recirculation line (see figure) and supplying the required flow rate $G_{RW(\text{mix})}$ of hot water through it, while:

$$G_{RW(prod)} = G_{RW(\text{mix})} + G_{RW(L)} = k_{RW} \cdot G_{RW(\text{mix})} = k_{RW} \cdot \alpha_{RW(HW)} \cdot G^{bh}_{HW(prod)}, \quad (8)$$

where $G_{RW(prod)}$, $G^{bh}_{HW(prod)}$ – the same as in equation (1); $G_{RW(\text{mix})}$ – consumption of hot water supplied (recirculated) from boiler units to their inlet; $G_{RW(L)}$ – consumption of recirculating hot water lost during its movement from boiler units to their inlet; k_{RW} – recirculating water loss factor; $\alpha_{RW(HW)}$ – the proportion of the recirculating hot water consumption in the hot water consumption generated by the boiler units.

The flow rate $G_{RW(\text{mix})}$ of recirculating hot water is found as a result of the joint solution of the equations of material and heat balances written for the mixing points B (see figure). Taking into account the equation of the material balance of boiler units:

$$G^{bh}_{IW(BU)} = G^{bh}_{HW(prod)}, \quad (9)$$

this solution gives:

$$G_{RW(\text{mix})} = \alpha_{RW(HW)} \cdot G^{bh}_{HW(prod)}, \quad (10)$$

where $G^{bh}_{IW(BU)}$ – the consumption of improved (heated) water entering the boiler unit inlet; $\alpha_{RW(HW)}$ – the same as in formula (8):

$$\alpha_{RW(HW)} = \frac{\tau_{1S(HW)} - \tau_{RW}}{\tau_{2(HW)} - \tau_{RW}}, \quad (11)$$

where τ_{RW}, $\tau_{2(HW)}$ – the same as in equation (5); $\tau_{1S(HW)}$ – design temperature of water entering boiler units, °C.

Initially, it is not possible to calculate the flow rates $G^{on}_{HW(prod)}$ and $G^{fu}_{HW(prod)}$ of hot water. We pre-set them using empirical expressions of the form:

$$G^{i}_{HW(prod)} = \alpha^{i}_{HW(HW)} \cdot G^{bh}_{HW(prod)}, \quad (12)$$

where the index i = *on*, *fu*; $\alpha^{on}_{HW(HW)}$, $\alpha^{fu}_{HW(HW)}$ – the share of hot water consumption, generated, respectively, to meet the own needs of the boiler house and the needs of the fuel economy, in the total consumption of hot water generated.

Solving equation (1) together with equations (2), (8) and (12), the flow rate $G^{bh}_{HW(prod)}$ of hot water produced by boiler units is found:

$$G^{bh}_{HW(prod)} = \frac{k^{hn}_{HW} \cdot \alpha_{HW(D)}}{1 - \alpha^{on}_{HW(HW)} - \alpha^{fu}_{HW(HW)} - k_{RW} \cdot \alpha_{RW(HW)}} \cdot G_d. \quad (13)$$

Hot water boiler units operate reliably in all modes if the flow rate of heated water circulating through them is constant [7 -10]. In this case, the thermal power of the boiler units (boiler room) in the range of outdoor temperature values $\tau_{obp} \ldots \tau_{doa}$ changes as a result of a change in the temperature $\tau_{2(HW)}$ of the hot water supplied by them. During this period of the year, all units installed in the boiler room work.

In the range of outdoor temperature values $\tau_{ob} \ldots \tau_{obp}$ and in the non-heating period of the year ($\tau_o > \tau_{ob}$), the temperature $\tau_{2(HW)}$ of hot water produced by boiler units does not change and is taken equal to its value at the break point of the temperature graph: $\tau_{2(HW)} = \tau_{2I(HW)} = \tau_{2(HW)}$, where $\tau_{2I(HW)}$ is the temperature of hot water supplied by boiler units at the break point of the temperature graph: (at $\tau_o = \tau_{ob}$), °C .

At outside air temperatures $\tau_o > \tau_{ob}$, the heat output of the boiler room is controlled by changing the number of operating boiler units - by starting or stopping a part of the boiler units. At the same time, through each operating unit, the estimated consumption of heated water is provided. Initially, at the design temperature $t_{\text{НО}}$ of the outside air, we find the flow rate $G^{bh}_{HW(prod)}$ of hot water produced by the boiler units at $\tau_o = \tau_{ob}$: $\tau_{2S(HW)} = \tau^*_{1S}$ and $\alpha_{HW(D)}$=1.

The resulting value of the flow rate of the generated hot water is calculated and is taken over the entire range of outdoor temperature values $\tau_{obp} \ldots \tau_{doa}$. Knowing the flow rate of the generated hot water, we calculate its temperature $\tau_{2(HW)}$ at the required values of the outdoor air temperature, including when $\tau_o = \tau_{ob}$. In the range of outdoor air temperature values $\tau_{ob} ... \tau_{ou}$ and in the non-heating period of the year, the flow rate $G^{bh}_{HW(prod)}$ of the generated hot water is found by the value of its temperature $\tau_{2u(HW)}$.

3. Investigation results

Solving equation (1) together with expressions (2), (8) and (12), and also taking into account relations (5) and (11), we determine the temperature of hot water produced by boiler units at $\tau_o = \tau_{obp} \ldots \tau_{doa}$:

$$\tau_{2(HW)} = \tau_{RW} + \frac{k^{hn}_{HW} \cdot G_D \cdot \left(\tau^*_1 - \tau_{RW}\right) + k_{RW} \cdot G^{bh}_{HW(prod)} \cdot \left(\tau_{1s(HW)} - \tau_{RW}\right)}{\left(1 - \alpha^{on}_{HW(HW)} - \alpha^{fu}_{HW(HW)}\right) \cdot G^{bh}_{HW(prod)}}. \quad (14)$$

The flow rate G_{IW} of the initial (raw) water and the flow rate $G^{\text{Д}}_{\text{ХОВ(ПСТ)}}$ of chemically treated water entering the deaerator are interconnected as follows:

$$G_{IW} = G_{IW(L)} + G_{CTW(L)} + G^D_{CTW(E)} = k_{IW} \cdot G^D_{CTW(E)}, \quad (15)$$

where G_{IW} is the flow rate of the source water entering the chemical water treatment area (CWTA); $G_{IW(L)}$ – the consumption of the source water lost as a result of its costs for the auxiliary needs of the chemical water treatment site and in the form of leaks; $G_{CTW(L)}$ – consumption of chemically treated water lost in the form of leaks during its movement to the deaerator; $G^D_{CTW(E)}$ – consumption of chemically treated water entering the deaerator; k_{IW} $k_{\text{ИВ}}$ is a coefficient that takes into account the loss of initial and purified water.

Consumption $G^D_{CTW(E)}$ of chemically treated water entering the deaerator and consumption $G^D_{HW(C)}$ of hot water consumed by the deaerator as heating coolant, we find as a result of the joint solution of the equations of material and heat balances of the deaerator. This solution gives:

$$G^D_{CTW(E)} = D_{LEAV} + G_{MW(S)}; \quad (16)$$

$$G^D_{HW(C)} = G_{MW(S)} \cdot \frac{\left(i_{DW} - i^d_{CTW}\right) + d \cdot \left(i_V - i^d_{CTW}\right)}{\left(i_{2(\text{HW})} - i_{DW}\right) - d \cdot \left(i_V - i^d_{CTW}\right)}, \quad (17)$$

where D_{LEAV} – flow rate of the steam-gas mixture (vapor) leaving the deaerator; $G_{MW(S)}$ – consumption of make-up water supplied by the deaerator to the heating network; d – specific consumption of vapor;

i_{DW}, i^{d}_{CTW}, $i_{2(\mathrm{HW})}$, i_{V} – specific enthalpies of deaerated, chemically purified, hot water and vapor, respectively.

Consumption $G_{MW(S)}$ of make-up water supplied by the deaerator is equal to:

$$G_{MW(S)} = k_{MW} \cdot (G_{D(L)} + G^{bh}_{HW(L)} + G_{RW(L)} + G^{on}_{RW(L)} + G^{fu}_{HW(L)}), \quad (18)$$

where k_{MW} – make-up water loss factor; $G^{i}_{j(L)}$ – consumption of water lost in meeting certain needs:

$$G^{i}_{j(L)} = \mu^{i}_{j} \cdot G^{i}_{j(C)} \left(j = d,\ HW,\ RW; i = hn,\ on,\ fu \right), \quad (19)$$

where μ^{i}_{j} – the share of the lost water consumption from the consumption $G^{i}_{j(MFR)}$ of the consumed water.

Thermal calculations of heaters 2 and 5, as well as vapor cooler 6, are carried out using their heat balance equations.

Having calculated the flow rates $G^{on}_{HW(prod)}$ and $G^{fu}_{HW(prod)}$, we check the accuracy of setting the values $\alpha^{on}_{HW(HW)}$ and $\alpha^{fu}_{HW(HW)}$ in expression (12).

Calculating the costs $G_{DW(E)}$, $G^{H}_{HW(E)}$ and $G^{CH}_{HW(E)}$, we check the accuracy of setting the values of the quantities $\alpha_{DW(WVR)}$, $\alpha^{H}_{HW(WVR)}$, $\alpha^{CH}_{HW(WVR)}$ in the formula (7).

The reliability of the performed calculations is estimated by checking the fulfillment of equality (9). Its right-hand side is calculated using expression (1). The calculation of the left side of equality (9) is carried out according to the equation:

$$G^{bh}_{RW(mix)} + G_{RW(mix)} = G^{bh}_{IW(BU)}, \quad (20)$$

where $G^{bh}_{RW(mix)}$ is the consumption of recovered water entering the boiler inlet:

$$G^{bh}_{RW(mix)} = G^{bh}_{d(E)} + G_{DW(E)} + G^{H}_{HW(E)} + G^{fu}_{HW(E)}. \quad (21)$$

4. Conclusion

The considered method of calculating the heating scheme of the boiler house was accepted for implementation by the Energy Management and Audit Laboratory of the State Institution "Oryol Regional Energy Saving Center" (State Institution "Orel RCE"). Its approbation took place for the first time in 2016, during an energy survey of a hot-water boiler house located on the territory of the FSBEI HE "OSU named after IS Turgenev". The results obtained using the developed methodology and instrumental examinations confirmed a high level of convergence of technological indicators (error - less than 5%). Taking into account the achieved positive effect, this technique is used for energy surveys of water heating boilers in Orel and the Oryol region, carried out by specialists of the State Institution "Orel RCE" with the participation of the developers of the technique. Its use makes it possible to increase the accuracy of the performed heat engineering calculations, and also provides a deeper understanding of the essence of heat engineering processes, and, as a result, simplifies the solution of the problem of optimizing the operating modes of boiler equipment [11, 12].

References

[1] SP 124.13330.2012 "SNiP 41-02-2003 Heating networks"
[2] Badaguev B T 2010 Steam and hot water boilers: operational safety: orders, instructions, magazines, regulations (Moscow: Alfa-Press) p 198
[3] Krol L B, Kelman G N 1970 Intermediate superheating of steam and its regulation in power units (Moscow: "Energy") p 320

[4] Kraush S A 2003 Heat-generating installations of heat supply systems: a textbook for university students (Tomsk: Tomsk State University of Architecture and Civil Engineering) p 161
[5] Gusev Yu L 2011 Basics of designing boiler plants 2nd ed. (Moscow: Stroyizdat) p 248
[6] Rivkin A S 2011 Thermal calculation of the boiler unit: textbook. Manual (Ivanovo: ISEU) p 144
[7] 1983 Boilers of low and medium power and combustion devices: 15-83. Industry catalog (Moscow: NIIEINFORMENERGOMASH) p 226
[8] Gubarev A V, Vasilchenko Yu V 2008 Heat generating installations. Part 1: tutorial (Belgorod: publishing house of BSTU im. V.G. Shukhova) p 162
[9] Shumilin E V, Psarov S A 2013 Thermal calculation of the boiler: workshop (Khabarovsk: Publishing house of the Pacific State University) p 78
[10] Danilov O L, Garyaev A B, Klimenko A V et al. 2011 Energy saving in heat power engineering and heat engineering: textbook for universities 2nd ed. (M.: Publishing house MEI) p 424
[11] Kachanov A N, Golenkov V A, Stepanov Yu S, Koroleva T G 2003 Methodology for conducting energy audits of public sector enterprises: a tutorial (Orel: Orel State Technical University) p 156
[12] Kachanov A N, Golenkov V A, Stepanov Yu S, Koroleva T G 2003 Methodology for designing energy surveys of public sector enterprises: Textbook (Orel: Orel State Technical University) p 156

IOP Conf. Series: Materials Science and Engineering 1211 (2022) 012009 doi:10.1088/1757-899X/1211/1/012009

A differential operation principle of the automatic bus transfer system pickup unit as a way of decreasing the emergency clearing time.[a]

Nikolay Danilov[1],Sergey Tsyruk[1], Alexandr Timonin[1] and Karam Sharafeddine[1,2]

[1] Moscow Power Engineering Institute, Moscow, Russian Federation
[2] Lebanese University,Beirut, Lebanon

Abstract. A proper choice of the design and operation algorithm of emergency control devices like high-speed bus transfer (HSBT) is only possible proceeding from a study and analysis of steady-state and transient processes in emergency modes of operation (short-circuit faults, power supply disconnection, or phase open-circuit fault). The numerical experiments for studying such modes that were carried out, using the Matlab Simulink software package, on the mathematical models of an industrial power supply system involving synchronous motors connected to it made it possible to synthesize a new differential HSBT pickup unit featuring a high-speed response to emergency events. In doing so, special attention was paid to an analysis of transient operation modes with the aim of minimizing the probability of false actuations. The obtained study results have found practical application in the HSBT devices installed at the facilities of PJSC MOSENERGO. The experience gained from the operation of a new device jointly with high-speed circuit breakers produced by the Tavrida-Elektrik state-owned corporation has demonstrated essential advantages in comparison with the conventional HSBT designs.

1. Introduction

The choice of the design of emergency control devices like fast bus transfer (FBT) arrangements for redundant power supply systems is primarily connected with the need to retain transient stability of synchronous motor load within the time interval taken to isolate an emergency and restore normal power supply.

The possibility of retaining the transient stability is determined by a certain limit time T_{crit} for which the synchronous motor (SM) load angle δ (Fig. 1a) [1] in the case of loss of power supply does not reach the critical value δ_{crit}, above which the motor's rotor will slip, thereby aggravating the resynchronization process (Fig. 1b).

The values of T_{crit} and δ_{crit} depend on the power supply system configuration and parameters, kind of emergency (short-circuit faults, tripping the input circuit breaker by relay protection, etc.), and also on the electric motor operation parameters (capacity utilization factor CUF and power factor cosφ) [2].

The conventional FBT arrangements are not used almost at all at enterprises having a continuous process cycle, including trunk pipeline transport, petrochemistry, etc. in view of their insufficient

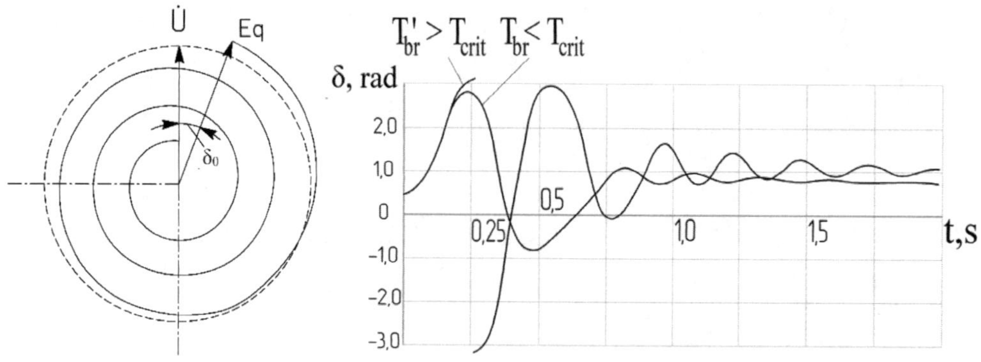

(a) (b)

Figure 1. Variation of the synchronous motor EMF Eq and angle δ in the case of loss of power supply (a) and synchronous motor resynchronization after power supply restoration (b).

operation cycle time equal to 30-40 ms has been developed, although---given the constantly increasing capacity of single electrical appliances and their high sensitivity to the power interruption time---even these operation times are criticized [3].

In redundant power supply systems that contain HSBT devices (Fig. 2), the total time taken to isolate emergencies is determined from the expression

$T_{int} = t_{HSBT} + t_{icb} + t_{tcb}$,

where t_{HSBT} is the HSBT response time; t_{icb} is the input circuit breaker opening time, and t_{tcb} is the tie circuit breaker closing time.

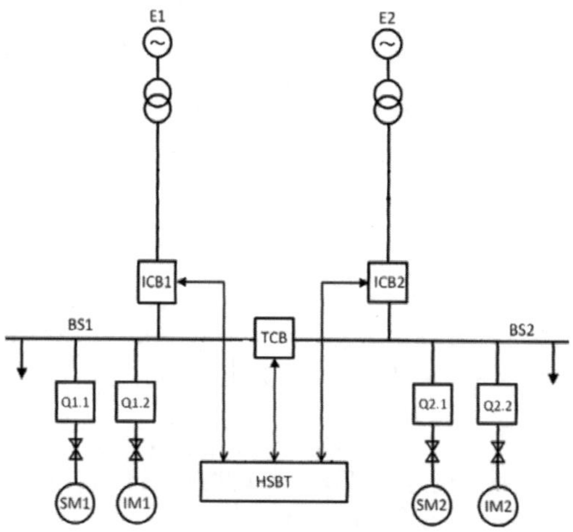

Figure 2. Typical power supply scheme with two independent inputs and two busbar sections.

The operation of circuit breakers accounts for the major part of the time in the HSBT cycle; improvement of their response speed can hardly be expected in view of potential switching over voltages and extremely high mechanical loads in their contact parts [4, 5]. In the HSBT algorithm, a shift is possible from the sequential command generation scenario [6], in which the commands for opening the input circuit breaker (ICB) and closing the tie circuit breaker (TCB) are produced simultaneously. This allows the HSBT cycle time to be decreased by 35-40% on the average. However, the statistics gained in the course of HSBT operation shows that improvement of the overall response speed entails a higher percentage of device false actuations, which points to an insufficiently mature structure of automatic controls and their operation algorithms.

The article studies the regularities and parameters characterizing the transients in power supply systems, based on which new principles for detecting the occurrence of emergencies are proposed. As an outcome, along with better HSBT response speed, the reliability of commands for device actuation has been improved.

2. Statement of the problem

An emergency at the switchgear input can be detected by using the following criteria:

- the input voltage is not in line with the required level;
- there is a deviation of the monitored input voltage frequency or phase;
- there are over currents in the load circuits;
- there are excessive levels of negative- or zero sequence voltages;
- the power flow changes its direction for the opposite.

The above-listed criteria are interrelated in some or other way, which adds some complexity to the analysis of operation parameters, because some of them are both enabling and disabling in nature. For example, a dip of input voltage may be a consequence of a short circuit fault both at the input and in the load circuits; however, in the second case the HSBT pickup unit must remain silent. As a second example, consider a situation in which a dip of input voltage occurred, but the value of zero sequence voltage exceeded its setpoint. On the one hand, it can be considered that the HSBT pickup unit should remain silent because there is a disabling $3U_0$ signal, but on the other hand, this situation may be a consequence of an open-circuit fault of one or two phases at the input, and a fast switchover to the standby power source must be carried out.

Let us find the response time of the HSBT pickup unit for the case of the HSBT conventional design [7]. We assume that the logical device controlling the pickup unit operation receives, in a real time mode, a continuous flow of discrete data characterizing the operating parameters. The initial values of operating parameters are periodic functions of time; therefore, for performing a fully valued analysis of each of the above-listed criteria, the time interval $T = 1/F_n$ (F_n = 50 Hz is the network frequency) is theoretically necessary; this interval is a "sliding window" [8], in which the sequence of stored data "moves" in time and is constantly updated with new values.

The parameter T will appear in each analytical expression for calculation of the effective and mean values, for the signal power, and for the frequency and phase. The frequency is calculated by analyzing discrete samples in the interval $T = 1/F_n$. For calculating the phase, the discrete Fourier transform (DFT) is used, in which the sampling frequency is selected so that the ratio $N = F_s/F_n$ (where F_s is the sampling frequency) is always an integer number. In our case, N is the "window" size the value of which will be a multiple of the network frequency F_n. The expressions for calculating the above-mentioned parameters can be presented as follows:

The effective voltage value for the cycle:

$$U_{eff} = \sqrt{\frac{1}{T}\int_0^T u^2 dt}; \quad (1)$$

The mean voltage value for the half-cycle:

$$U_m = \frac{2}{T}\int_0^{T/2} u dt; \quad (2)$$

The mean power (the analytical expression):

$$P_m(\varphi) = \frac{2}{T}\int_0^{T/2} \frac{UI}{2}(\cos(\varphi) - \cos(2\omega t + \varphi))dt; \quad (3)$$

The mean power (for discrete systems):

$$P_m(\varphi) = \frac{2}{T}\int_0^{T/2} \frac{UI}{2}(\cos(\varphi) - \cos(2\omega t + \varphi))dt; \quad (3)$$

$$Px_m = \frac{\sum_{k=0}^{m/2} u_k i_k}{m/2}; \quad (4)$$

$$\text{DFT: } X_k = \sum_{k=0}^{k=N-1} u_k e^{-\frac{2\pi i}{N}kn}; \quad (5)$$

where k = 0…N-1, $N = F_s/F_n$.

It should be noted that in contrast to the other parameters, the mean power is calculated by a factor of two faster, because formula (3) for calculating it contains the cosine of a double argument, which in certain situations involving predominantly motor load makes it possible to achieve a factor of two higher response speed. In some operation modes, the algorithm response speed can be increased by shifting the actuation setpoint closer to the normal operation mode. In this case, however, it is obvious that as a result of narrowing the relay hysteresis loop, the use of such method may lead to unexpected actuations under normal network operation conditions. Thus, in the conventional systems the HSBT response speed cannot be increased without additional processing of the monitored signals.

3. The pickup unit differential operation principle

Before to consider the differential pickup unit [9] in detail, the following note should be made: a step change of the frequency or phase is a mathematical abstraction, because in reality in digital devices these quantities are always calculated inside the "sliding window"; therefore, they vary quite smoothly within the interval $T = 1/F_n$. Thus, for example, the system will "see" a step change of the voltage phase only when this fragment of digital sequence will fully turn to be in the data window, i.e., in N cycles after the event onset (Fig. 3). In the case of a step frequency change, all samples relating to the previous mode must leave the data window, which is also possible in N cycles (here and henceforth, the following values were chosen as input data: F_s = 2400 Hz, and N = 48).

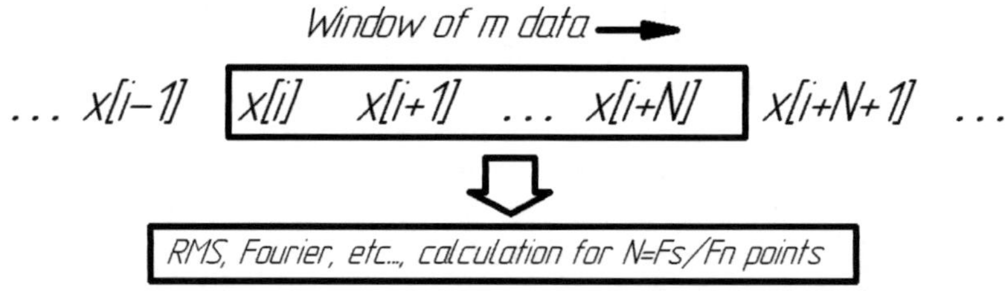

Figure 3. Calculation using the "sliding window" method (F_s is the sampling frequency, and F_n is the network frequency).

For revealing step changes in the function being studied, the analysis of the effective (RMS) value in the interval T can be replaced by comparison of the current quantity with a reference function taken from the preceding or earlier cycles. This operation is similar in its sense to differentiation, but is simpler to implement in comparison with it and is free from uncertainties at the discontinuity points of the function being analyzed. The output signal produced by the pickup unit constructed in this manner is proportional to the input signal variation rate and is effective only for detecting fast changes at the device inlet. With such an arrangement, the response time for discrete signals corresponds to one sampling step. For signals having a comparatively slow variation rate (with respect to their previous valued in the interval T), the pickup unit effectiveness decreases proportionally; therefore, it is better to analyze these signals using the conventional methods. Thus, it can be concluded that a pickup unit that complies with all conditions in terms of response speed and emergency detection efficiency should in our opinion have, by analogy with PID automatic control [10, 11], the integral-differential structure shown in Fig. 4.

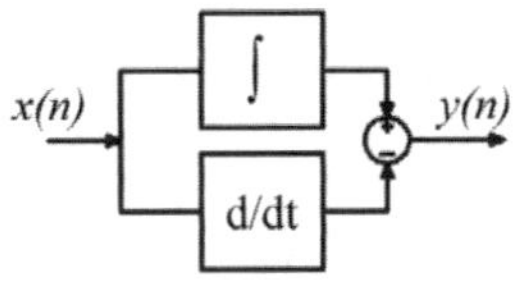

Figure 4. Structure of the integral-differential pickup unit.

In should be borne in mind that unnecessarily high response speed of the pickup unit may result in undesirable actuations when short-term voltage dips occur, which do not have an essential influence on the power supply quality. Thus, if the pickup unit comes in action in response to a 1.5-ms long voltage dip, this may result in a power supply interruption with the duration t ≥ 12 ms depending on the types of the used circuit breakers and control method [12]. Therefore, to decrease false actuations, we will consider the revealed faults with the duration $t_f \leq 2$ ms as interference, and the other faults will be regarded as potentially hazardous from the viewpoint of the possibility to evolve into a real emergency [1].

Instrument current transformers (CTs) and voltage transformers (VTs) serve as data sources for the HSBT system. It should be borne in mind that instrument transformers provide the specified accuracy class only for steady state values. In transient modes, the instrument transformer behavior is not standardized as a consequence of the nonlinear properties of the transformer steel under the nonzero pre-emergency mode conditions. If a nonzero current flow through the CT primary winding, the transformer will store the following energy in each half-cycle:

$$E = \frac{L_\mu I_\mu^{\ 2}}{2}; \tag{6}$$

where L_μ is the magnetization inductance, and I_μ is the magnetization current amplitude?

If the current in the primary circuit is abruptly disconnected at the time moment with E ≠ 0, the current through the transformer secondary winding will not become zero until the transformer energy stored in the previous cycle is fully transferred. If an inrush current occurs, the current increase rate in the CT secondary winding will be determined by the leakage inductances L_s1, L_s2 of the CT equivalent circuit, which in turn depend on the CT design features.

Voltage transformers may physically consist of several interconnected single-phase transformers or a combination of a three-leg E-type transformer and a single-leg shell-type transformer. Besides, anti-resonance-type VTs include additional zero-sequence transformers (ZSTs), which may have various parameters of their magnetic system depending on the particular design version. Situations should be considered for VTs when one or two phase voltages disappear in the network. In this case, a mode similar to a single-phase ground fault (SPGF) with a simultaneous occurrence of a zero sequence voltage U_0 across the VT open delta occurs. The voltage at the open delta output is given by the following expression:

$$U_0 = \frac{\overrightarrow{U_a} + \overrightarrow{U_b} + \overrightarrow{U_c}}{\sqrt{3}}, \tag{7}$$

where $\overrightarrow{U_a}, \overrightarrow{U_b}, \overrightarrow{U_c}$ are the reduced phase voltage vectors.

With a deep voltage dip [13], e.g., in the C phase ($U_c = 0$), in accordance with (7) and the cosine theorem, at the open delta output we obtain

$$U_0 = \frac{\sqrt{U_a^{\ 2} + U_b^{\ 2} + 2U_a U_b \cos(\alpha)}}{\sqrt{3}} = \frac{U}{\sqrt{3}}, \tag{8}$$

where U is the effective value of the reduced phase voltage.

Since the phase voltage at the instrument VT input is always reduced to the value $100/\sqrt{3}$, at any phase voltage dip moment, we obtain across the VT open delta the value $U_0 = 100/3$ V. From (7) it follows that the same voltage appears at the above-mentioned output when two phase voltages disappear simultaneously. For finding the output voltage in the case of a SPGF of any phase, in (8) we substitute, instead of phase voltages, line-to-line voltages with the angle $\alpha = 60^0$, as a result of which we obtain $U_0 = 100$ V.

This example shows the following:

- the transient triggered as a consequence of a dip of any phase voltage will depend on the pre-emergency mode and also on the type and internal parameters of the used VT;
- any imbalance of the input voltages after completion of the transient in the VT results in the occurrence of the zero sequence voltage U_0, which can be mistakenly interpreted as an SPGF.

In view of the reasoning presented above, the data obtained from instrument transformers for 1-2 ms from the transient onset moment should be regarded as unreliable; otherwise, this may provoke false operation of the pickup unit. Besides, for correct operation of the pickup unit, the SPGF mode recording setpoint should be above 100/3 V.

4. Numerical experiments and validation of the results

For checking how correctly the HSBT control algorithm operates in emergencies differing from one another in the kind and occurrence place, a pickup unit analysis mathematical model was developed, which was implemented in the Matlab (Simulink) environment [14] (Fig. 5).

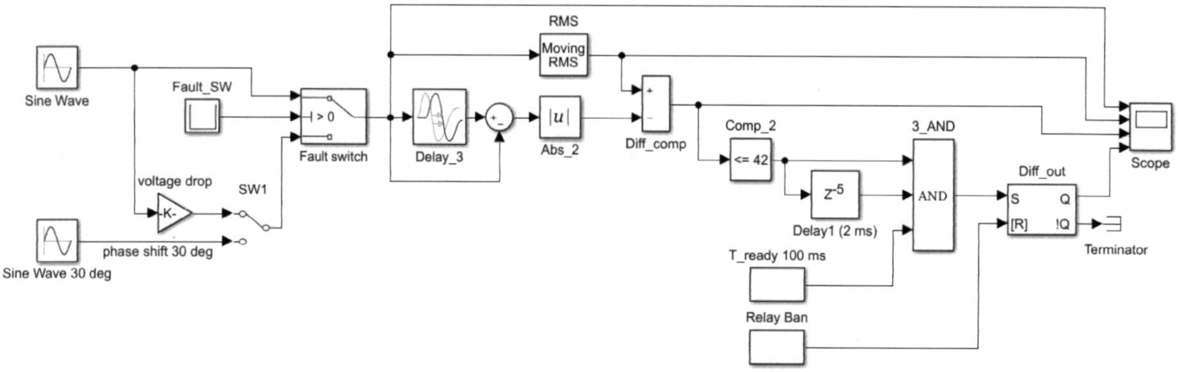

Figure 5. Differential pickup unit analysis model.

For analyzing the model behavior, the following disturbances are applied to its input: in the first case, it is a sinusoidal signal 60 sin(ωt) with its amplitude decreased by a factor of two at the time moment 0.155 ms, and in the second case, a step change of the phase by 30° occurs in the input signal under the same conditions. The input signal type is selected by means of the SW1 switch. The actuation threshold in response to the minimal voltage is set equal to 30% of the nominal phase voltage ($100/\sqrt{3}$ V). The model uses the external HSBT actuation readiness signal (T_Ready) and the actuation banning signal from relay protection and automatic control devices (Relay_Ban). The model output value is the following logical values at the RS flip-flop output (Diff_Out): 1 when the pickup unit came in action, and 0 when the pickup unit is in the waiting mode.

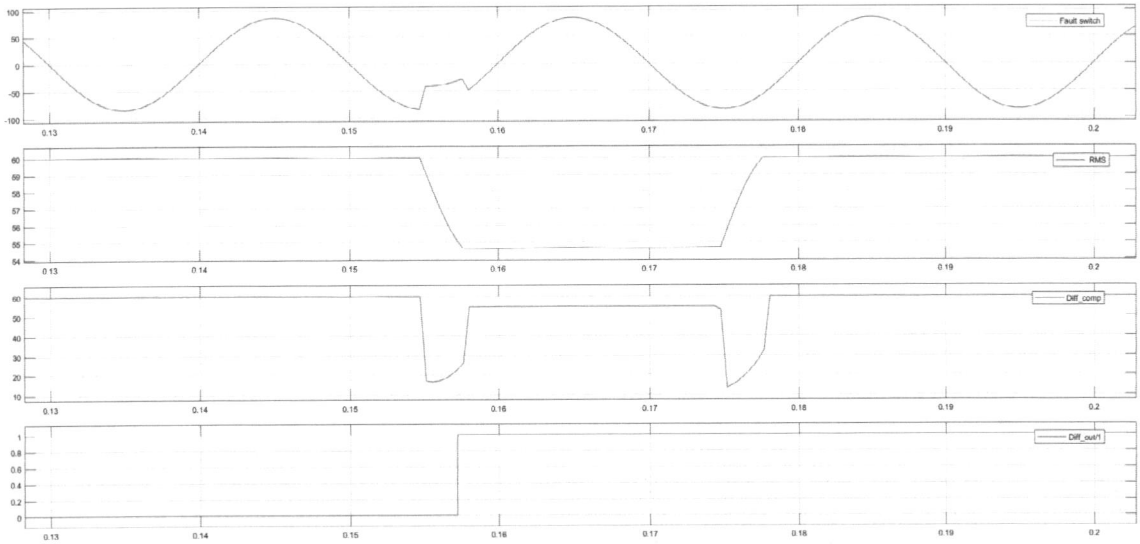

Figure 6. Simulation of a short-term voltage dip at the power supply input.

It can be seen from Fig. 6 that the 50% voltage dip with a duration of 3 ms, which occurred at the time moment t = 0.155 ms causes the voltage at the output of the integral type pickup unit to decrease to 54.6 V, whereas the voltage at the output of the integral-differential pickup unit decreases to 20 V, thereby allowing the emergency process onset to be detected in a clear-cut manner.

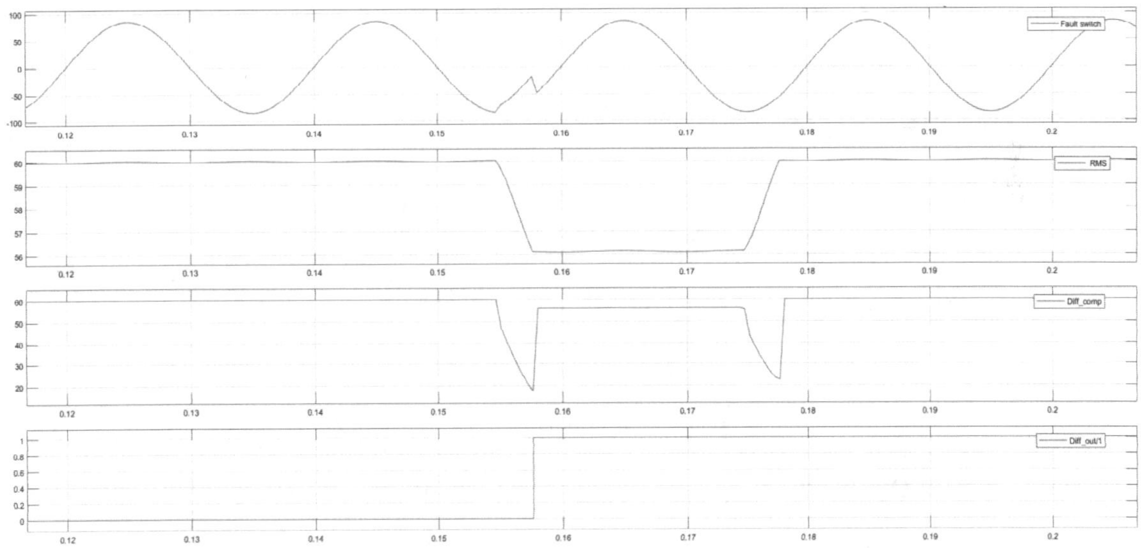

Figure 7. Simulation of a short-term phase shift at the power supply input.

In Fig. 7, a stepped change of the phase by 30° with a duration of 3 ms occurs at the time moment t = 0.155 ms, causing the voltage at the integral type pickup unit output to decrease to 56.2 V, whereas the voltage at the output of the integral-differential pickup unit decreased to 19.6 V, thereby also allowing the emergency process onset to be reliably detected.

5. Conclusions

The key specific feature of the proposed pickup unit is the versatility of its operation algorithm, due to which it is able to analyze all potentially hazardous situations irrespectively of the load type and nature. Another advantage lies in the simplicity of its implementation, which does not involve the need to perform trigonometrical calculations of the current values of the phase angle δ and frequency F for detecting the emergency onset.

Further studies of methods for minimizing the operation time of emergency control devices like an HSBT seem to be advisable toward improving the design of switching apparatuses (circuit breakers) in terms of achieving their faster operation. Preference in this regard should be given to vacuum and SF_6 circuit breakers.

6. References

[1] Gamazin S I, Stavtsev V A, Tsyruk S A *Transients in industrial power supply systems triggered by electric motor load*, A Monograph, MEI, 1997.

[2] Danilov N V, Tsyruk S A, Ryzhkova Ye N. Studying the power supply critical interruption time as a high-speed bus transfer success criterion. *2021 Int. Conf. on Industrial Engineering, Applications and Manufacturing (ICIEAM-2021).* DOI: 10.1109/ICIEAM51226.2021.9446425

[3] Pupin V, Safonov D, Fedorov O Modeling and research of operating modes of oil and gas pumping stations to determine the requirements for automation devices. *International Ural Conf. on Electrical Power Engineering, UralCon.* 2020. 9216310. DOI: 10.1109/UralCon49858.2020.9216310.7

[4] Gandhare W Z, Jaonjare R S Protection of utility motors during switching. *IEEE International Conference on Computer Communication and Control*, IC4 2015. DOI: 10.1109/IC4.2015.7375618

[5] Gandhare W Z, Jaonjare R S Protection of utility motors during switching. *2015 Int. Conference on Computing, Communication and Security,* ICCCS 2015. DOI: 10.1109/CCCS.2015.7374208

[6] https://new.siemens.com/global/en/products/energy/energy-automation-and-smart-grid/protection-relays-and-control/siprotec-4/high-speed-busbar-transfer-device/high-speed-busbar-transfer-siprotec-7vu68.html.

[7] Danilov N V, Danilov R N, Tsyruk S A, Gamazin, S I, Petrov V S *RF Patent No. 2662639, "A method and device for implementing high-speed bus transfer*", 2018.

[8] Jwo D-J, Chang W-Y, Wu I-H Windowing techniques, the welch method for improvement of power spectrum estimation. 2021 Computers, Materials and Continua. **67(3)**, pp. 3983-4003

[9] Danilov N V, Tsyruk S A and Petrov V S *RF Patent 2717236 "A method and device for implementing high-speed bus transfer"*, 2020

[10] Zhmud V, Dimitrov L, Nosek J Effective algorithm to design pid-regulator by numerical optimization. international multi-conference on industrial engineering and modern technologies, *FarEastCon 2020.* DOI: 10.1109/FarEastCon50210.2020.9271402.

[11] Miller Y V, Sablina G V Calculation of PID-controller parameters (2020). *Automatics & Software Enginery*, **1 (31)**, pp. 148-153.

[12] ABB VM1 Installation and Service Instructions Manual. https://www.manualslib.com/manual/1335536/Abb-Vm1.html#manual

[13] *IEC TR 61000-2-8:2002.* Electromagnetic Compatibility (EMC) - Part 2-8: Environment - Voltage Dips and Short Interruptions on Public Electric Power Supply Systems with Statistical Measurement Rsults.

[14] Demiroren A, Zeynelgil H L Modelling and simulation of synchronous machine transient analysis using SIMULINK. *Int. Journal of Electrical Engineering Education*, **39 (4)**, 2002. pp. 337-346. DOI: 10.7227/IJEEE.39.4.

IOP Conf. Series: Materials Science and Engineering 1211 (2022) 012010 doi:10.1088/1757-899X/1211/1/012010

Method for assessing the state and residual life of the insulation based on analysis of the time of discharge of the capacity of electric cables

Ivan Maksimovich Kazymov[1], Boris Sergeevich Kompaneets[1]

[1]Altai State Technical University named I. I. Polzunov, Barnaul, Russian Federation

Abstract. As of today, the problem of the lack of non-destructive methods for assessing the state and determining the residual life of the insulation of electrical cables is relevant for cable lines in operation, and especially for lines that have exhausted their service life. The research presented in the framework of this article is aimed at creating a non-destructive technique for assessing the state and residual life of electrical cables insulation. The study was carried out using computer simulation methods and using the basic theoretical laws of physics and electrical engineering. A method for assessing the condition and residual life of electrical cable insulation has been developed, which is based on the analysis of the discharge time of the cable capacity. The applicability of the method is determined, instructions are given on its applicability and the analysis of the output data. The obtained results can be used by power grid companies and industrial enterprises to analyze the state of the insulation of electrical cables.

1. Introduction

Assessment of the state of insulation of electrical cables and, in particular, determination of its residual resource is currently one of the most important areas of the main activity of electric grid companies due to the need to schedule repairs and maintenance of electrical equipment, as well as assessing the need for repair a this or that cable line.

Modern methods of assessing the suitability of an electric cable for a voltage over 1000 V in the Russian Federation are based on testing the insulation of a cable with increased voltage according to the Rules of technical operation of electrical installations of consumers [1], in the world – on the basis of similar regulatory documents. However, the application of increased voltage to the insulation of electrical cables in some cases can lead to the appearance and development of insulation defects that were absent before the application of the test voltage. At the same time, there are cases when the insulation of electrical cables, being under an normal operating voltage limited by the permissible deviation limits, provided the possibility of normal operation of the cable line. However, if insulation defects, allowing the line to continue operating, appear as a result of tests with increased voltage, the condition of the line will worsen with the development of the arisen defects even during normal operation conditions and, ultimately, this will lead to the failure of the cable line. The mentioned processes are covered in detail in a number of Russian [2-4] and foreign scientific studies [5-7].

Also, when conducting tests with increased voltage, minor insulation defects can develop into a full-fledged irreversible breakdown and lead to the need to repair the cable line.

Thus, high-voltage insulation testing is a destructive testing technique.

To date, many studies have been conducted and published on this topic [8-10], various options for assessing the state of cable insulation and residual life are offered, but no effective methodology for evaluating a cable line based on a examining of the cable's own capacity has yet been proposed.

In this regard, there is a need to create a methodology for assessing the condition and residual insulation life of an electrical cable of a non-destructive nature, based on research on the capacity of the cable line, which, however, provides reliable data on the residual insulation life of the cable.

IOP Conf. Series: Materials Science and Engineering 1211 (2022) 012010 doi:10.1088/1757-899X/1211/1/012010

2. Problem statement

The methodology for assessing the condition and residual insulation life of electrical cables is being developed to assess the insulation of three-wire cables at a voltage of 6/10 kV with paper soaked insulation, armored and unarmored.

The developed methodology for assessing the condition and residual insulation life of electrical cables should have the following properties:

- be a non-destructive testing technique;
- allow a comparative analysis of the insulation condition of the wires of one cable or different cables;
- have strict algorithmic calculations;
- be based on the basic laws of physics and electrical engineering, have a physical meaning;
- to ensure the reliability of the obtained results sufficient for practical use.

The application of the developed methodology should be carried out using a technical equipment available to electrotechnical laboratories.

3. Materials and methods

The methodology for assessing the condition and residual insulation resource of electric cables is based on the analysis of the discharge time of the capacity of the cable line, the charge of which is carried out during tests with increased voltage. At the same time, the test voltage can be arbitrary: from the required by the standards of testing for a cable of this voltage class and type of insulation to the one selected according to local conditions to ensure the non-destructive nature of the research.

When testing a cable with increased voltage according to the traditional wiring diagram for connecting the test unit (Figure 1), the cable represents a capacitor, due to the presence of distributed capacitance between the wires and on the cable sheath. The leakage current flows through this capacitor due to the finite insulation resistance of the cable, as well as due to the effects of absorption and polarization. But the contribution the effects of absorption and polarization, however, is leveled off during the test due to the prolonged application of the test voltage (according to the standards of testing of electrical equipment in the current edition – on practice, the duration of the application of the test voltage is 5 minutes). Thus, with prolonged exposure to the test voltage, the leakage current is characterized only by the state of insulation and its active resistance.

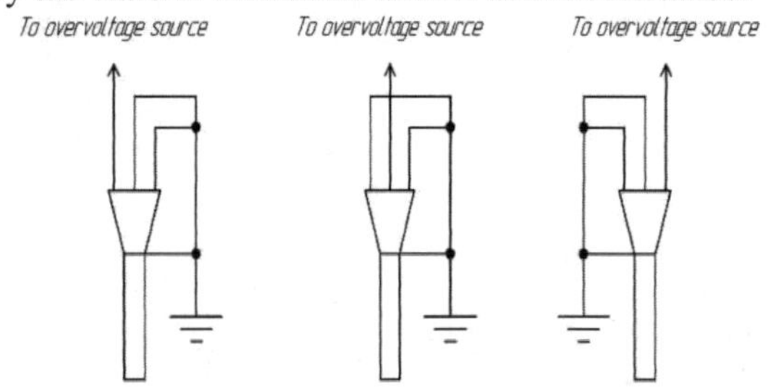

Figure 1. Typical wiring diagram for connecting the test unit.

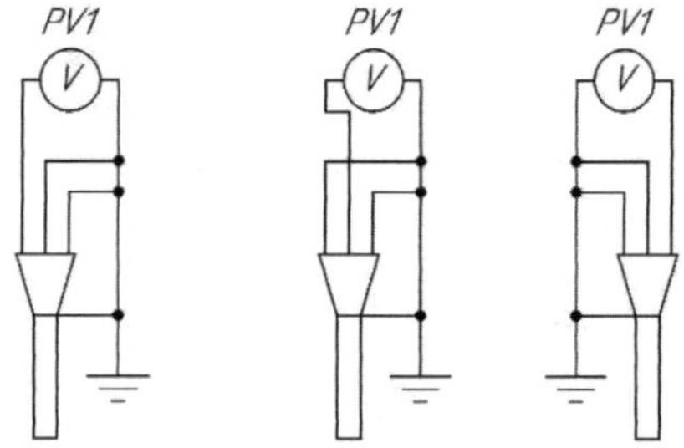

Figure 2. Wiring diagram of connecting a voltmeter for each cable wire.

When the test voltage is stopped and the test unit is disconnected from the cable line, the tested cable wire will remain under full test voltage relative to the other two wires and the grounded surfaces at the initial moment of time. As a result, the process of discharging the capacitance in the cable begins along the only available circuit – through the active insulation resistance.

Obviously, the better the insulation condition, the higher its resistance and the longer the process of discharging the cable capacity will be. This process can be controlled using a chronometer and a high-precision voltmeter, which allows measuring voltage in range from zero to full test value. In this case, it is required to write down and construct a discharge graph in time-voltage coordinates. The wiring diagram of connecting the voltmeter is shown in Figure 2. The view of the discharge graph for the general case is shown in Figure 3.

IOP Conf. Series: Materials Science and Engineering 1211 (2022) 012010 doi:10.1088/1757-899X/1211/1/012010

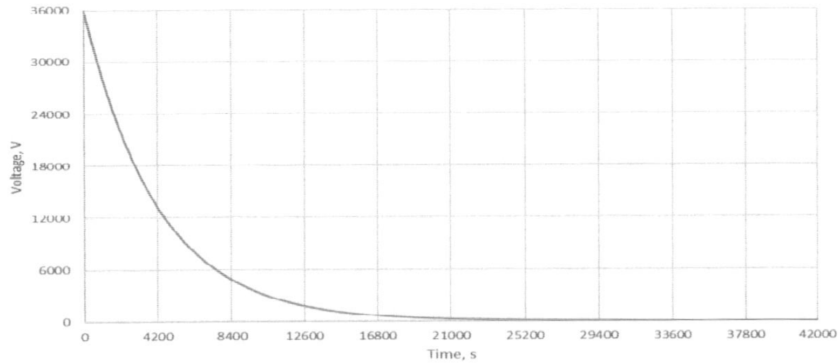

Figure 3. Cable discharge graph (initial voltage: 36 kV; specific capacity: 0.42 μF/km, line length: 10 km; insulation resistance: 1000 MOhm).

At the end of the construction of the discharge graph (Figure 3), it is required to determine the time constant based on the characteristic points of the discharge graph and their correspondence to the typical discharge graph of the capacitor – the voltage reduction to a level of 36.8% of the initial should occur 3 times faster than the reduction to a level of 5.0% of the initial and 5 times faster than 0.7% of the initial. If the actual data significantly differ from the assumed ones, it is concluded that there are insulation defects in cable, the influence of which weakens with a decrease of the voltage value.

If the discharge graph corresponds to a typical one for a capacitor, it is required to determine the insulation resistance based on the obtained value of the time constant and the cable capacity according to the formula (1):

$$R_{is} = \frac{\tau}{C_c}, \tag{1}$$

where τ – discharge time constant, seconds;

C_c – cable capacity, μF.

In this case, the cable capacity can be calculated by the formula (2):

$$C_c = c_0 \cdot l_c, \tag{2}$$

where c_0 – specific capacity of the cable (reference data), μF/km;

l_c – cable length (design and/or operational data), km.

It is worth noting that the results obtained by formula (1) R_{is} are presented in megaohms (MOhm) and do not require additional conversion to the insulation resistance measurement units used during testing.

Based on the analysis of the insulation resistance obtained and the correspondence of the discharge graph typical for a capacitor, it can be concluded about state of the insulation a separate cable wire. When carrying out similar actions with other two wires of the cable, a comparative analysis of the insulation condition can be carried out. For a general assessment of the insulation condition of an electric cable, it is recommended to take the smallest of the received resistances as the final insulation resistance. If there are significant differences in the insulation resistance of different wires or in the deviation of one of the discharge graph from the typical one, it is concluded that there are insulation defects. Suitability for further use of cable line in this case is determined only on the basis of the results of tests with increased voltage with a standard level of test voltage.

To obtain more accurate data, the nature of the discharge process flow can be studied for voltmeter wiring diagrams other than those shown in Figure 2. Alternative voltmeter wiring diagrams are shown in Figures 4-7.

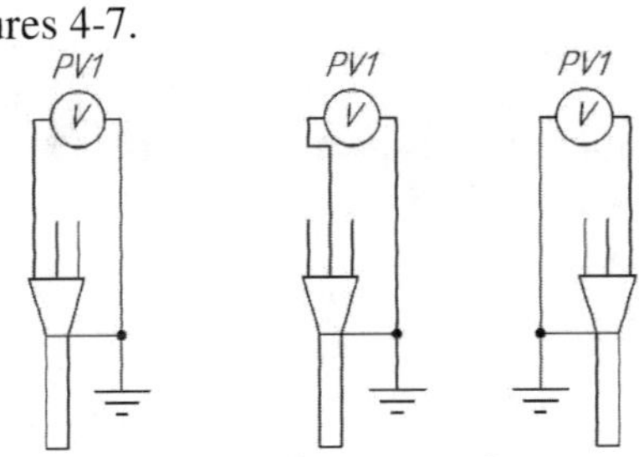

Figure 4. Wiring diagram of connecting a

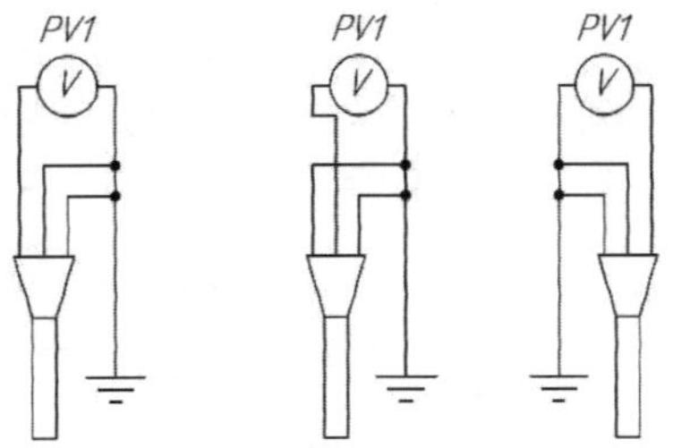

Figure 5. Wiring diagram of connecting a

IOP Conf. Series: Materials Science and Engineering 1211 (2022) 012010 doi:10.1088/1757-899X/1211/1/012010

voltmeter for each cable wire.

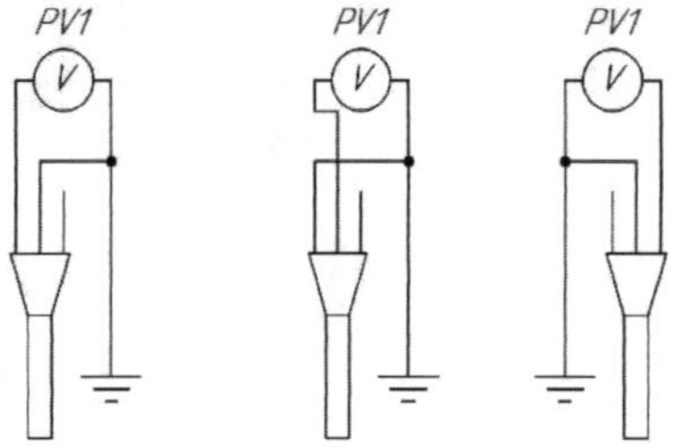

Figure 6. Wiring diagram of connecting a voltmeter for each cable wire.

voltmeter for each cable wire.

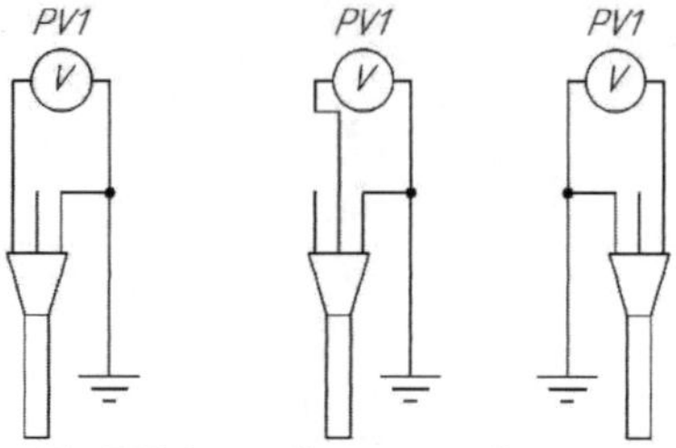

Figure 7. Wiring diagram of connecting a voltmeter for each cable wire.

It is obvious that in the vast majority of cases in practice, for power cables with a voltage above 1000 V, it is not possible to monitor the discharge process according to the schemes shown in Figures 5-7, due to the absence of a real possibility to ensure reliable insulation of the cable sheath from grounded surfaces along the entire length of the cable due to the fact that the cable sheath is attached to grounded surfaces at the end cutting points and structurally has no insulation from armor and/or protective cover designed for the operating voltage of the cable.

In all cases, the discharge graph must correspond to the typical one for the capacitor, however, the discharge time constant will be different for each of the connection schemes due to changes in cable capacitance (not all cable capacity are involved during discharge).

An assessment of changing the discharge time constant, provided that the insulation resistance remains unchanged, will allow us to estimate the change in the actual capacity of the cable under various switching schemes. The resulting capacitance values must satisfy the following expression (3):

$$C_c = C_0 + 2 \cdot C_p, \quad (3)$$

where C_c – capacitance between one wire and to the metal sheath and between same wire and other two wires, μF;

C_0 – capacity between wire and sheath, μF;

C_p – capacity between the wires, μF.

According to reference data, the capacity between wire and sheath and the capacity between the wires are related by the expression (4):

$$C_p = 0{,}27 \cdot C_0, \quad (4)$$

where C_0 – capacity between wire and sheath, μF;

C_p –capacity between the wires, μF.

4. Results and Discussion

The insulation resistance value can be obtained by formula (1) both using the calculated cable capacity determined by formula (2) and using the actually measured capacity. The processes of measuring capacity are not considered in this research.

To carry out a comprehensive assessment of the cable insulation condition, it is proposed to analyze the discharge time of the cable capacity with different initial voltages, which will allow to confirm or deny a significant change in the shape of the discharge graph or the value of the discharge time constant (and, consequently, the insulation resistance) due to changes in the initial discharge conditions. In the case of significant changes in the nature of the discharge process flow (example in Figure 8), for various initial voltages, it is concluded that there are insulation defects that express themselves with an increase of the voltage applied to the insulation. In this case, the defective part of the insulation can be detected by examining the insulation condition according to the various presented wiring diagrams of connecting a voltmeter. In this case, it becomes possible to determine whether a part of the insulation of the wire from the sheath or part of the insulation between the wires is defective. An example is shown in Figure 9 (an example when the insulation resistance values obtained are different for different switching schemes).

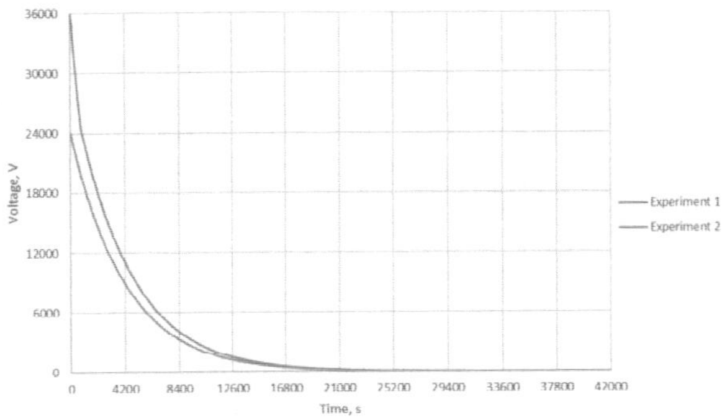

Figure 8. A comparative picture of the nature of the discharge process flow at different initial voltages.

It is easy to notice that in Figure 8, the discharge of the cable in Experiment 1 at the voltage range from the initial 36 kV to a value close to 24 kV proceeds much faster than at the voltage range from 24 kV to full discharge for each of the experiments. Obtaining such a comparative picture indicates the presence of insulation defects, the influence of which begins to affect when a constant voltage exceeding 24 kV is applied to the insulation of the cable.

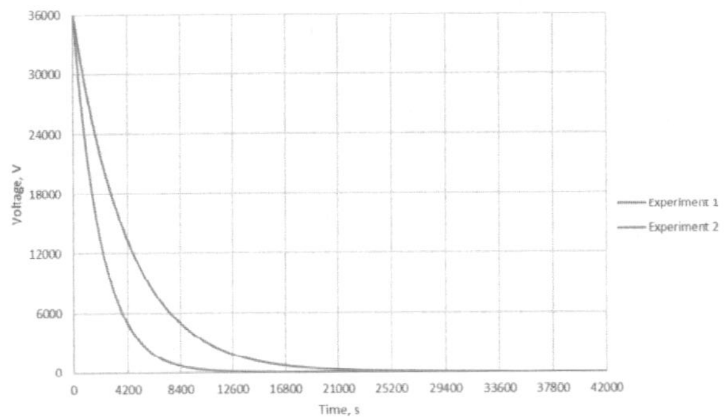

Figure 9. A comparative picture of the nature of the discharge process flow with different voltmeter connection schemes (various controlled insulation resistances).

It is observed from Figure 9 that the discharge in experiment 2 proceeds much faster, which, given the same initial voltage, indicates a lower insulation resistance compared to that studied in Experiment 1. Similar differences for different insulation resistances (for example, for insulation resistance between wires and between the wire and the sheath) are the expected picture due to the known design features of the electric cable. At the same time, significant deviations of discharge graphs showing significant deviations of insulation resistance values for theoretically predicted identical insulation resistance values (for example, different insulation resistance values between different wires and the sheath or between different wires) clearly indicate the presence of insulation defects and even allow localizing the damaged zone of insulation.

4.1 A model of spreading of the leakage current through the insulation from the test wire to other wires and the sheath.

Graphically, the discharge of charge from the charged core to the grounded parts of the cable with a standard wiring diagram for increased voltage testing (Figure 1) is schematically shown in Figure 10.

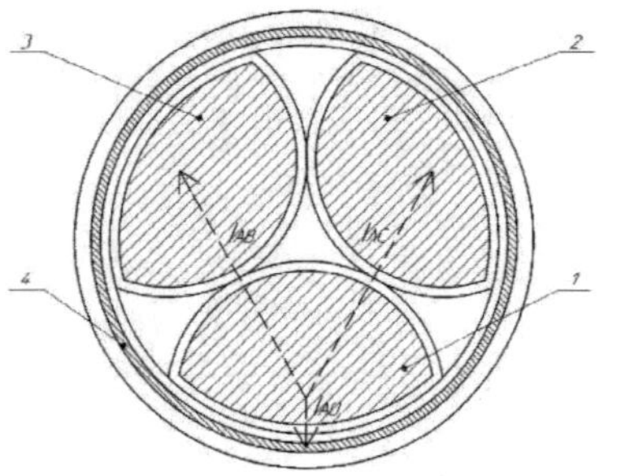

Figure 10. A schematic representation of the process of charge draining from the charged core to the grounded parts of the cable at the end of the increased voltage testing process for the standard wiring diagram for wire A of the cable (I_{AB}, I_{AC}, I_{A0} – charge drain currents from the charged wire A to the wires B, C and the sheath, respectively; 1 - wire A of the cable; 2 - wire B of the cable; 3 – wire C of the cable; 4 - cable sheath).

It is easy to notice that the cable assembled according to the test circuit (Figure 1) at the end of the overvoltage set and its holding for a long enough time is a composite capacitor for the flow of electric current due to the termination of the effects of absorption and polarization. At the same time, this capacitor consists of three capacitors physically connected in parallel and having one common positive voltage plates (negative voltage plates are interconnected and grounded). The schematic diagram of the cable for the flow of electric current is shown in Figure 11. The only way to discharge such a capacitor is the transfer of charge between the plates through an insulation layer having a finite resistance.

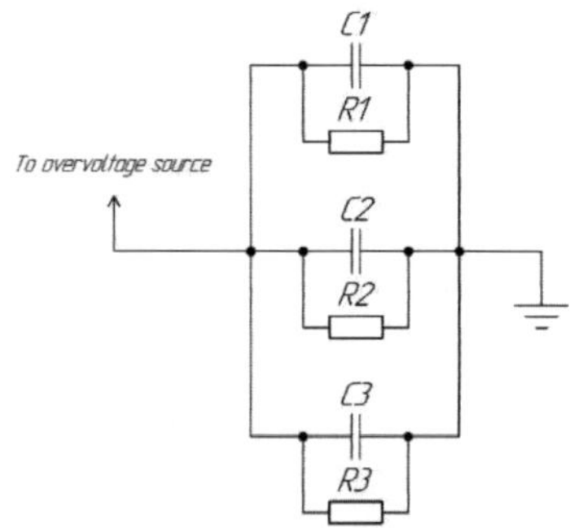

Figure 11. Circuit diagram of the cable for the flow of electric current through the insulation in the increased voltage test mode with and at its end (C_1, C_2 – the capacitance between the test wire and another wire; C_3 – the capacitance between the test wire and the cable sheath; R_1, R_2 – insulation resistance between the tested wire and another wire; R_3 – insulation resistance between the tested wire and the cable sheath).

4.2 A model for comparing discharge processes.

Two different methods can be used to compare the two discharge processes:

- comparison of two discharges according to a different wiring diagram at the same initial voltage (at the moment of coincidence of the current voltages) – in this case, the discharge time and cable capacity will be different;
- comparison of two discharges according to a different wiring diagram with the same initial voltage for an arbitrary discharge time – in this case, the current voltage and cable capacity will be different.

For the first method, considering that the discharge equation is generally described by the dependence (5), it seems possible to find using equation (6) the capacitances ratio of the tested cable for different discharge schemes.

$$U(t) = U_0 \cdot e^{-\frac{t}{R \cdot C}}, \qquad (5)$$

where U_0 – initial voltage, V;

t – discharge time, s;

R – insulation resistance, MOhm.

C – capacity of the cable, μF.

$$U_0 \cdot e^{-\frac{t_1}{R \cdot C_1}} = U_0 \cdot e^{-\frac{t_2}{R \cdot C_2}}, \qquad (6)$$

where t_1, t_2 – discharge time spent to achieve the same current voltage in various discharge experiments, s;

R – insulation resistance, MOhm.

C_1, C_2 – cable capacity in various discharge experiments, μF.

By the simplest arithmetic operations, an equation of the following form (7) can be obtained:

$$\frac{t_1}{t_2} = \frac{C_1}{C_2}. \tag{7}$$

Provided that the discharge time ratio (t_1 to t_2), and the assumed capacitances ratio (C_1 and C_2) to the capacitance of the cable C_c are known, the correspondence of the theoretical dependencies to the experimental ones can be checked and the assumption of the good insulation condition can be confirmed or disproved, since in the case of damaged sections, the discharge time will be significantly reduced and, therefore, provided that the insulation resistances R in formula (6) remain unchanged, the ratio of capacities C_1 to C_2 (obtained by calculation from equation (7)) will not coincide with the theoretically expected one. If the insulation is in good condition, the discharge process will be same according to any of the proposed discharge schemes.

For comparison of two discharge processes using the second described method will initially need to express the dependence (5) through the discharge time t according to the formula (8).

$$t = -R \cdot C \cdot \ln(\frac{U(t)}{U_0}). \tag{8}$$

In this case, to compare two discharge process using the second described method will need to create an equation of the form (9):

$$\frac{C_1}{C_2} = \log_{\frac{U(t)_1}{U_0}} \frac{U(t)_2}{U_0}, \tag{9}$$

where $U(t)_1, U(t)_2$ – current voltage value on the cable during discharge time t in various discharge experiments, V;

U_0 – initial voltage, V.

Provided that the voltages $U(t)_1$, $U(t)_2$ and U_0 are known, the process of assessing the insulation state is similar to the one described above for the first comparison method.

4.3 Determination the insulation resistance between the wires and the sheath.

To determine the insulation resistance between the wires and the sheath, it is necessary to connect the cable wires according to the wiring diagram shown in Figure 12 and connect an increased voltage source. At the end of the full test voltage set process and its holding for a time period required by the standards of testing of electrical equipment, it is required to connect measuring devices – voltmeters according to the diagram shown in Figure 13. In this case, the cable will consist of three capacitors, the discharge of each of which will be produced due to the finite insulation resistance between the wires and the sheath. So, when obtaining the discharge graphs of each of the wires to the sheath, conclusions about the state of insulation of the wires from the sheath can be made both in relative (the difference in the nature of the discharges between the different wires) and in absolute (mathematically calculated values of insulation resistance).

During the experiment, it is possible to obtain numerical values of insulation resistances R_{A0}, R_{B0}, R_{C0} by calculation on the basis of empirically constructed discharge dependences of the various cable capacitances for different resistances.

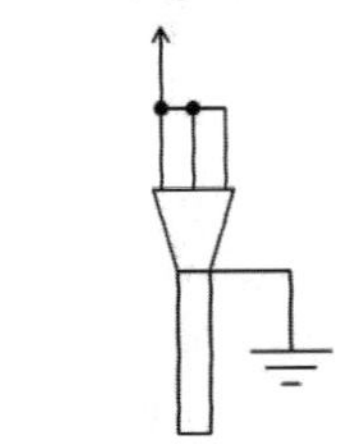

Figure 12. Cable connection diagram for determining the insulation resistance of the cable wires relative to the sheath.

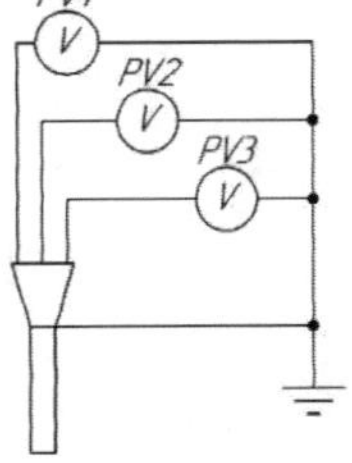

Figure 13. Voltmeters connection diagram for determining the insulation resistance between the cable wires and the sheath.

4.4 Determination of insulation resistance between cable wires.

It is obvious that during the analysis of the discharge graph of the cable according to the traditional diagram of the increased voltage testing, the final value of the insulation resistance obtained as a result of calculations consists of the insulation resistances between the wire and the sheath and between the wire and other two wires connected in parallel. This resistance for the phase A cable wire is described by the expression (10).

$$R_{\Sigma A} = \frac{R_{A0} \cdot R_{AB} \cdot R_{AC}}{R_{AB} \cdot R_{AC} + R_{A0} \cdot R_{AC} + R_{A0} \cdot R_{AB}}, \tag{10}$$

where R_{A0} – insulation resistance between the phase A cable wire and sheath, MOhm;
R_{AB} – insulation resistance between the phase A wire and phase B wire, MOhm;
R_{AC} – insulation resistance between the phase A wire and phase C wire, MOhm.

It should be noted that the resistances $R_{\Sigma A}$, $R_{\Sigma B}$, $R_{\Sigma C}$ in this article are assumed to be obtained by formula (1). Expressions similar to expression (10) can be made to describe the insulation resistance of the wires of other phases. In this case, a system of three equations (11) containing three unknowns can be compiled, and each equation will contain two unknowns.

$$\begin{cases} \frac{1}{R_{AB}} = X - \frac{1}{R_{AC}} \\ \frac{1}{R_{AB}} = Y - \frac{1}{R_{BC}}, \\ \frac{1}{R_{AC}} = Z - \frac{1}{R_{BC}} \end{cases} \tag{11}$$

where X – numerical value of the expression $X = \frac{1}{R_{\Sigma A}} - \frac{1}{R_{A0}}$, MOhm;
Y – numerical value of the expression $Y = \frac{1}{R_{\Sigma B}} - \frac{1}{R_{B0}}$, MOhm;
Z – numerical value of the expression $Z = \frac{1}{R_{\Sigma C}} - \frac{1}{R_{C0}}$, MOhm.

The solution of this system is three expressions (12, 13, 14) that uniquely determine the insulation resistance between the wires based on certain values of the insulation resistance between the wire and sheath and the final insulation resistance between the wire and sheath and between wire and two other wires.

$$R_{AB} = \frac{2}{X+Y-Z}; \tag{12}$$

$$R_{BC} = \frac{2}{Y-X-Z}; \tag{13}$$

$$R_{AC} = \frac{2}{Z-Y+X}. \tag{14}$$

Therefore, within the proposed method, it becomes possible to determine separately the values of insulation resistances between the wires and between the wires and the sheath, which will allow a comprehensive analysis of the insulation condition and assess its residual resource.

4.5 Assessment of the insulation condition.

It is proposed to evaluate the insulation condition based on the following parameters:
- absolute value of insulation resistance;
- coefficient of deviation of calculated values of insulation resistance of various cable wires;
- coefficient of deviation of the ratio of capacities between the wires and between the wires and sheath.

The numerical values of the parameters proposed as criteria for the insulation condition are given in Table 1 for power electrical cables with paper insulation at a voltage from 6 to 35 kV.

The numerical parameters of the absolute values of insulation resistance in Table 1 for a "Satisfactory" insulation condition are given with the assumption that the values of leakage currents of electric cables correspond to the normalized values for power cables in the Rules of Technical Operation of electrical installations of consumers [1], namely, the maximum permissible leakage current for a power cable. The values of the absolute values of the insulation resistance for the "Good" and "Excellent" insulation condition are given with the assumption that the leakage current is reduced by two and four times, respectively. At the same time, based on the standard cable design, the insulation resistances between the core and the shell and between the cores are assumed to differ twice.

The coefficient of deviation of the insulation resistance values shows the uneven wear of the cable insulation and may indicate the presence of local insulation damage, which may be caused by various factors, such as:

- systematic overload of one of the cable wires with increased current loads, leading to overheating of the insulation of the wire – in this case, the insulation both between this wire and other wires, and between this wire and the sheath will be significantly reduced, and the insulation resistance between this wire and the sheath will be reduced more due to the fact that the insulation resistance between the wires includes two independent, electrically connected in series, insulation resistances of the wires;
- deformation of the cable, leading to a partial violation of the insulation layer – in this case, mostly reduced an insulation resistance between one or more wires and the sheath relative to the insulation resistance to the sheath of intact wires.

The coefficient of deviation of the cable capacitance values demonstrates the difference between the empirically obtained values of the cable capacitance ratios between the wires and between the wire and the sheath from the theoretically predicted values. This difference can be caused by various reasons:

- measurement error – due to the fact that the value of the capacitance and the calculation of its ratios are performed by indirect methods and depend on the accuracy of measurements of other quantities (including non-electrical), it is difficult to perform accurate measurements of the cable capacity, however, the value of a relative error of 2% is quite an achievable level of measurement accuracy;
- deterioration of insulation– due to the fact that the process of determining the capacitance ratios is based on analyzing the nature of the discharge process with different connection diagrams, the resulting value of the capacitance ratio depends on the discharge time to the same value of the current voltage, as well as on the expected ratio of insulation resistances with different connection diagrams: in this case, the deviation of the true resistance ratio from the theoretically predicted value will affect the resulting value of the capacitance ratios and indicate the presence of insulation damage in the cable.

Table 1. Numerical values of parameters for assessing the state of insulation of electrical cables.

№	Rated cable voltage, kV	Rated test voltage, kV	Insulation resistance between the wires, MOhm	Insulation resistance between the wire and the sheath, MOm	Coefficient of deviation of insulation resistance values, %	Coefficient of deviation of capacitance ratios, %	Insulation condition
1	6	36	720	360	from 2,5 to 10,0	from 5,0 to 10,0	Satisfactory
2	6	36	1 440	720	from 1,0 to 2,5	from 2,0 to 5,0	Good
3	6	36	2 880	1 440	below 1,0	below 2,0	Excellent
4	10	60	480	240	from 2,5 to 10,0	from 5,0 to 10,0	Satisfactory
5	10	60	960	480	from 1,0 to 2,5	from 2,0 to 5,0	Good
6	10	60	1 920	960	below 1,0	below 2,0	Excellent
7	20	100	268	134	from 2,5 to 10,0	from 5,0 to 10,0	Satisfactory
8	20	100	536	268	from 1,0 to 2,5	from 2,0 to 5,0	Good
9	20	100	1 072	536	below 1,0	below 2,0	Excellent
10	35	175	280	140	from 2,5 to 10,0	from 5,0 to 10,0	Satisfactory
11	35	175	560	280	from 1,0 to 2,5	from 2,0 to 5,0	Good
12	35	175	1 120	560	below 1,0	below 2,0	Excellent

Moreover, it should be noted that exceeding the value of the actual capacity of the cable relative to the value of its nominal (design) capacity with taking into account the actual length of the cable, should be regarded as a positive factor positively characterizing the state of insulation and favorably affecting the residual resource of the cable insulation.

4.6 Assessment of the residual insulation resource.

It is proposed to evaluate the residual insulation resource on the basis of an assessment of the insulation condition as follows:
- all indicators of the insulation condition indicate an «Excellent» insulation condition – the residual resource is assumed to be comparable to the passport service life, there are no contraindications to testing the cable with a normative value of increased voltage;
- at least one of the indicators indicates a «Good» insulation condition (the others are the same or higher) – the residual resource is assumed to be within 10-15 years; there are no contraindications to testing the cable with a normative value of increased voltage;
- at least one of the indicators indicates a «Satisfactory» insulation condition (the others are the same or higher) – the residual resource is assumed to be within 5 years; it is not recommended to test the cable with a normative value of increased voltage;
- at least one of the indicators indicates an unsatisfactory state of insulation – the insulation resource has been exhausted, or a spot repair of the cable is required (in this case, the insulation should be burned and search for the place of damage should be carried out).

5. Conclusion

The non-destructive nature of the proposed method for monitoring the state of cable insulation consists in the use of a reduced test voltage coupled with the use of the described method for assessing the state of insulation based on the analysis of the discharge time of the cable capacity due to the fact that the test voltage lowered to values determined by local conditions allows testing the cable insulation for withstanding the operating voltage with a given margin factor, after which the insulation condition is assessed according to the proposed method.

Thus, cable insulation does not experience excessive loads during testing and measurements, and the genesis of previously non-existent and the development of existing minor defects is not provoked in the insulation body. However, with sufficient correctness and accuracy for practical purposes, the detection of existing insulation defects is carried out on the basis of indirect measurements. And each type of detected defect can be digitized and comparatively assessed in relation to other cable wires or similar cables in the electrical network of the operated object(s). Within the proposed methodology, are determined both the values of insulation resistances and the values of the capacitances between the wires and other wires and between wires and sheath, as well as their mutual ratios.

The correctness of the methodology for assessing the insulation state based on the analysis of the discharge time of the cable capacity in various diagrams is confirmed by comparing theoretical and experimental data on the discharge of the capacitor on the load (the final active insulation resistance).

Using the proposed methodology for assessing the state of insulation and residual life allows, together with using a test voltage value determined based on local conditions, to extend the useful resource of the cable due to the absence of cases of applying a voltage to the insulation that is many times higher than its normative operational value.

This article presents a method for determining the insulation resistance of a cable based on experimentally obtained data and cable passport values, which makes it possible to evaluate the insulation resistance value within the of high-voltage direct current cable insulation tests.

Also this article presents a model of spreading of the leakage current through the insulation, on the basis of which are given recommendations for comparing the nature of the discharge (two methods) with different cable connection diagrams. The obtained data make it possible to assess the state of insulation in linguistic categories and numerically predict the residual resource of cable insulation.

The theoretical value of this study consists in the development of a methodology for determining the insulation resistance of a cable based on the analysis of the discharge time of the cable capacity, as well as in the development of a model of spreading of the leakage current through the insulation in a three-wire cable with a conductive sheath.

The practical value of this study consists in the possibility of its application during planned testing of cable lines in operating companies in order to conduct non-destructive testing of the insulation condition and to schedule repairs and replacement of cable lines.

References

[1] Rules for the technical operation of electrical installations of consumers: approved by Order of the Ministry of Energy of Russia dated 13.01.2003 N 6 (as amended on 13.09.2018) "On approval of the Rules for the technical operation of electrical installations of consumers" (Registered in the Ministry of Justice of Russia on 22.01.2003 N 4145): put into effect from 01.07.2003 (In Russian).

[2] Vysogorets S. P., Nikonov D. I., Nazarychev A. N., Tadjibaev A. I. Problems of technical diagnostics of 6 (10) kV cable lines with paper-impregnated insulation in modern conditions. Proc. of the Int. Scientific Seminar named Yu.N. Rudenko, Syktyvkar, July 04–09, 2016 «Methodological issues of research on the reliability of large energy systems. – 2016. – P. 180–186 (In Russian).

[3] Biryulin V. I., Kudelina D. V., Larin O. M. Method of monitoring the state of insulation of cable lines. *Bulletin of the Voronezh State Agrarian University*. – 2020. – **Vol. 3.** – P. 38–45. DOI: 10.17238/issn2071-2243.2020.3.38 (In Russian).

[4] Novikova F. Sh. Zh. Da K., Mironova I. S., Chernov K. V., Maksyutov I. N. Assessment of the technical condition of high-voltage cables based on the frequency diagnostic method. *Theoretical and applied aspects of modern science.* – 2014. – Vol. 6-3. – P. 88–91 (In Russian).

[5] Granizo R., Platero C. A., Guerrero J. M. HVDC Ground Fault Detection Method Based on the Shield Leakage Current Measurement for Long Cable Energization. Proc. of 2020 IEEE Int. Conf. on Environment and Electrical Engineering and 2020 IEEE Industrial and Commercial Power Systems Europe (EEEIC / I&CPS Europe). 2020. P. 1–5. DOI: 10.1109/EEEIC/ICPSEurope49358.2020.9160778.

[6] Zhou L., Cao J., Wang S., Liu L., Yang Y., Jiang H. Comprehensive State Evaluation of High Voltage Cable Based on Multi-state Variables Characteristics and Variation Law. *Gaodianya Jishu/High Voltage Engineering.* 2019. **Vol. 45(12).** P. 3954–3953. DOI: 10.13336/j.1003-6520.hve.20190424004.

[7] Korzhov A. V., Gorshkov K. E., Korzhova M. E. Neural Networks in Monitoring and Forecast Reliability Cable Network. Proc. of 2018 Global Smart Industry Conf. (GloSIC). 2018. P. 1–6. DOI: 10.1109/GloSIC.2018.8570153.

[8] Zhang H., Mu H., Zou X., Zhang D., Zhang G., Xie C. A Novel Method for Identifying Cable Defect and Improving Location Accuracy Based on Frequency Domain Reflectometry. Proc. of 2021 IEEE 4th International Electrical and Energy Conf. (CIEEC). 2021. P. 1–5. DOI: 10.1109/CIEEC50170.2021.9510503.

[9] Montanari G. C., Seri P., Bononi S. F., Albertini M. Partial Discharge Behavior and Accelerated Aging Upon Repetitive DC Cable Energization and Voltage Supply Polarity Inversion. *IEEE Transactions on Power Delivery.* 2021. **Vol. 36(2).** P. 578–586. DOI: 10.1109/TPWRD.2020.2984766.

[10 Siodla K., Rakowska A., Noske S. The Proposal of a New Tool for Condition Assessment of Medium Voltage Power Cable Lines. *Energies.* 2021. **Vol. 14(14).** P. 4116. DOI: 10.3390/en14144116.

Information about the authors:

Ivan M. Kazymov, postgraduate at the Department of «Electrification of production and life»;
Phone: +7-(913)-093-52-10;
eLibrary SPIN: 8464-5810; ORCID: 0000-0001-6873-0315; Scopus ID: 57209794071;
E-mail: bahek1995@mail.ru.

Boris S. Kompaneets, Ph. D. (Engineering), Docent, Head of the Department of «Electrification of production and life»;
eLibrary SPIN: 7371-4290; ORCID: 0000-0001-5980-1230; Scopus ID: 57209796863;
E-mail: kompbs@mail.ru.

Modelling of Solar-Diesel Hybrid Power Plant

A G Vaskov, N Y Mozder and A F Narynbaev

Dept. of Hydropower and Renewable Energy National Research University "Moscow Power Engineering Institute" Moscow, Russia

Abstract. The article highlights the problems of distributed energy generation and focuses on solar-diesel hybrid power plant modelling and optimization. Designing power systems based on renewable energy sources includes a very relevant task of building mathematical models of such systems and their elements. The article presents an approach and definition of mathematical models describing photovoltaic-diesel (PV-D) hybrid power system elements used in decision making processes as a part of PV-D operation control. An overview of PV module output power, performance and temperature models is given. Along with the analysis of the specific fuel consumption dependencies on the operating power of the diesel generator, an example of diesel power plant unit commitment is shown.

1. Introduction

On the territory of the Russian Federation, there are large number of districts, where energy supply is distributed and normally is carried out by diesel power plants. Therewith isolated power systems are characterized by a number of problems:

- Outmoded generating equipment with high rate of specific fuel consumption.
- Negative environmental impact.
- High electricity generation costs.

According to [1], requirements for energy supply systems efficiency [2] are formulated as follows: the energy efficiency of power plants must be ensured "by implementing legal, organizational, scientific, industrial, technical and economic measures aimed at the efficient use of fuel and energy resources and integration of renewable energy sources in economic turnover".

The implementation of several dozens of projects for the modernization of isolated power systems on the Far Eastern Federal District territory [4] showed that the construction of power stations combining diesel-driven generation and a power plant based on an unstable renewably energy source makes it possible to build energy complexes with good technical and economic characteristics. This is achieved by a decrease in specific fuel equivalent consumption per unit of generated electrical energy for the entire power plant [2] (henceforth specific consumption).

Getting lower specific consumption rate becomes possible when optimizing the design and operational characteristic of a hybrid power system, which requires verified mathematical models of its elements:

- a model of primary renewable energy source,

- a technical model (a description of specific technical solutions allowing to implement the power supply function),
- an economical model (an evaluation of technical and organizational solutions in economic (financial) indicators).

For the tasks of designing and controlling PV-D hybrid power systems, these mathematical models should also have contradictory properties: on the one hand, they should give the most accurate description of environmental conditions and technical features; on the other hand, to be fairly simple and clearly evident to use. Also, different groups of mathematical models are used to model the design and operation of PV-D systems. The purpose of the first ones is to calculate the steady-state operating modes of such systems (most commonly modeling hourly average performance indicators), the second ones are to be used for simulating instantaneous operating modes. An overview of methods for modeling steady-state PV-D hybrid power systems modes are presented below.

2. Mathematical Modelling of PV Power Plant

2.1. Solar PV module output power model

The output power P_{PV} of a single PV module is determined by its geometric dimensions, efficiency and solar irradiance incidence in one of two ways:

- the output power of a PV module depends on solar irradiance [5]:

$$P_{PV}(t) = P_{RPV} \cdot \left(\frac{G(t)}{G_{ref}} \right) \cdot \eta_{PV} \quad (1)$$

where P_{RPV} is PV module rated power (W); G is plane of array (POA) irradiance (W/m^2); G_{ref} is solar irradiance at reference conditions (usually set as 1000 W/m^2) ; η_{PV} denotes PV module efficiency.

- In [6] for calculating PV module output power an expression is used that takes its dimensions into account:

$$P_{PV}(t) = \eta_{PV} \cdot A \cdot G(t) \quad (2)$$

where A is a surface of a single PV module used in the system (m^2).

2.2. Evaluation of PV module efficiency

PV module efficiency is a value that depends on the environmental parameters, since ambient temperature affects the main electrical quantities, such as voltage and current of photovoltaic module. In [7] temperature impact on PV module efficiency is presented, where following nonlinear relationship is proposed:

$$\eta_{PV} = \eta_{ref} \left(1 - \beta_{ref} \cdot \left(T_c - T_{ref} \right) + \gamma \log_{10} G \right) \quad (3)$$

where η_{ref} is PV module efficiency at standard test conditions (STC); β_{ref} is dimensionless temperature coefficient; T_{ref} is reference temperature (K); T_c is PV cell temperature (K); γ is solar irradiance correction factor.

Solar irradiance correction factor usually reduces to zero, so that the expression is equated to following:

$$\eta_{PV} = \eta_{ref} \left(1 - \beta_{ref} \cdot \left(T_c - T_{ref} \right) \right) \quad (4)$$

Equation (4) is a traditional linear expression of PV module efficiency [8]. In recent studies this model is most commonly applied to describe the efficiency of a solar PV module [9, 10].

As it was stated in [11], the nonlinear PV module efficiency model is not appropriate, since the efficiency values are higher than those obtained experimentally in all cases. It is proposed to calculate the efficiency of PV module using the linear model.

The advantage of applying this model is that it has simple and easily accessible required parameters and can be used for various climatic conditions.

2.3. Temperature coefficient

The value of the temperature coefficient β is normally given by the manufacturer in PV module specifications. However, this value can be calculated using the following formula [12]:

$$\beta = \frac{1}{T_0 - T_{ref}} \tag{5}$$

where T_0 is material-dependent temperature at which the electrical efficiency of a PV module drops to zero [13].

2.4. PV cell temperature model

The correlations suggested in the literature usually express T_c as a function of the corresponding meteorological parameters, such as ambient temperature T_a, local wind speed V_w and POA solar irradiance G.

The simplest equation for the steady state operation mode relates T_c to the ambient temperature and solar irradiance as follows:

$$T_c = T_a + k \cdot G \tag{6}$$

where k is the Ross coefficient, which represents PV cell temperature increase with increasing irradiance.

Linear expression (6) is valid for no load operation and does not take the wind effect into consideration. The main difficulty of using it is to determine the Ross coefficient. Table 1 shows the values of k parameter for various PV modules mounting types.

Table 1. Empiric Ross Coefficients [14]

Mounting type	k, (K·м^2/W)
Free standing	0.021
Flat roof	0.026
Sloped roof (well cooled)	0.020
Sloped roof (not so well cooled)	0.034
Sloped roof (highly integrated, poorly ventilated)	0.056
Facade integrated (transparent PV cells)	0.046
Facade integrated (opaque PV cells)	0.054

The most common way to determine PV cell temperature is to use normal operating cell temperature (NOCT). The PV cell temperature depends on the ambient temperature T_a and irradiance G and can be described by following equation [15]:

$$T_c = T_a + (NOCT - 20) \cdot \frac{G}{800} \tag{7}$$

For taking the wind speed effects into account two models are utilized. They consider wind convection of free standing PV arrays for wind speeds greater than 0 m/s [16]:

$$T_c = T_a + \left(\frac{0.25}{5.7 + 3.8V_f} \right) \cdot G \tag{8}$$

$$T_c = T_a + \left(\frac{0.32}{8.91 + 2.0V_f} \right) \cdot G \tag{9}$$

where V_f is free flow wind speed upwind of the PV array, which can be calculated as $V_f = \frac{V_w + 0.5}{0.68}$ .and V_w .is the wind speed at the PV array installation geographical point

Equations (8) and (9) can be used only for a case of PV arrays mounted on free standing frames.

However, this limitation may be eliminated by using the Ross coefficient, which would consider PV modules mounting type. Thus, (9) is transformed into the following:

$$T_c = T_a + \omega \cdot \left(\frac{0.32}{8.91 + 2.0V_f} \right) \cdot G \tag{10}$$

Table 2 shows the values of ω dimensionless empiric coefficient.

Another model includes the wind speed [17]:

$$T_c = T_a + \frac{(\tau\alpha) \cdot G_t}{U_0 + U_1 V} \tag{11}$$

where V is the wind speed, τ and α are optical transmission coefficient and absorption coefficient respectively, U_0 is a coefficient describing irradiance effect on the PV module temperature, while U_0 describes the cooling by the wind. These coefficients are determined according to the location.

The following model uses empirical coefficients considering PV module types and its mounting methods. The suggested values of the empirical coefficients *a* and *b* are summarized in Table 3 [18].

The PV cell temperature model has the following form [18]:

$$T_c = T_a + G \cdot \exp(a + bV) \tag{12}$$

In one of the latest studies the following temperature model has been derived [19]:

Table 2. ω Empiric Coefficient Values [16]

Mounting type	ω
Free standing	1.0
Flat roof	1.2
Sloped roof	1.8
Facade integrated	2.4

Table 3. *a,b* Coefficients Values

PV module type	Mounting type	*a*	*b*
Glass/cell/glass	Free standing	-3.47	-0.059
Glass/cell/glass	Rooftop	-2.98	-0.047
Glass/cell/polymer	Free standing	-3.56	-0.075
Glass/cell/polymer	Integrated	-2.81	-0.046
Polymer/film/steel	Free standing	-3.58	-0.113

$$T_c = \frac{U_L T_a + \left[(\tau\alpha) - \eta_{stc}\left(1-\beta T_{ref}\right)\left(1+\gamma_{Pmp}\cdot\ln\left(\frac{G_g}{G_0}\right)\right)\right]\cdot G_g}{U_L + \eta_{stc}\beta\left(1+\gamma_{Pmp}\cdot\ln\left(\frac{G_g}{G_0}\right)\right)\cdot G_g} \tag{13}$$

where U_L is the heat loss coefficient; G_g is solar irradiance at the PV array installation geographical point, η_{stc} and β are the module efficiency and the temperature coefficient, respectively, at STC; γ_{Pmp} is a dimensionless coefficient which is between 0.03 and 0.12 for single crystalline silicon.

The heat loss coefficient U_L is calculated as follows:

$$U_L = U_{L0} + U_{L1}\cdot V \tag{14}$$

where U_{L0} and U_{L1} are coefficients considering heat losses by convection and radiation, which are calculated for a specific location.

The above models were tested; the results are summarized in table 4 [19].

Table 4. Experimental Comparison of PV Cell Temperature Models

PV cell temperature model	R^2 (%)	RMSE (°C)
(7)	93.97	3.85
(8)	98.30	1.24
(9)	98.47	1.22
(11)	98.61	1.27
(12)	97.44	2.03
(13)	98.77	1.13

It can be concluded from the Table IV that these models have satisfactory accuracy and can be applied to determine the PV cell temperature. The results show that the inclusion of the wind parameter makes PV module temperature calculations more accurate.

3. Mathematical Modelling of Diesel Power Plant

The simplest approach to describing a diesel power plant (DPP) is to represent it as a voltage source [20]. The disadvantage of this approach, undoubtedly, is the representation of the DPP as an equivalent power source, which is equal to the sum of the installed capacities of all diesel-engine generators installed at the DPP.

From the design perspective, under this approach, the determining criterion for choosing a diesel generator is the maximum peak of load curve, i.e. the DPP output power must not be lower than the total peak load [21].

The restrictions on minimal $P_{DPP}^{\min}$ and maximum $P_{DPP}^{\max}$ operating power of the DPP are also considered:

$$P_{DPP}^{\min} \le P_{DPP}(t) \le P_{DPP}^{\max} \tag{15}$$

In this case, the DPP operating power is determined on the basis of the system power balance equation:

$$P_{DPP}(t) = P_{load}(t) - P_{PV}(t) \tag{16}$$

where $P_{load}(t)$ is hourly average power consumption (kW).

In case of diesel generators reduction in DPP it is not possible to determine the values of the specific fuel consumption of individual diesel generators due to differences in their individual energy characteristics. For a more accurate modeling of the specific fuel consumption, it is necessary to consider the typical sizes and the number of diesel generators installed at the DPP, as well as their load conditions and wear.

It is known [22] that in order to ensure stable and reliable operation of diesel generators, it is necessary to distribute active power between them in proportion to their nominal power.

Optimal loading of diesel generators is provided on the basis of individual dependences of specific fuel consumption on the diesel generator operating power (Figure 1).

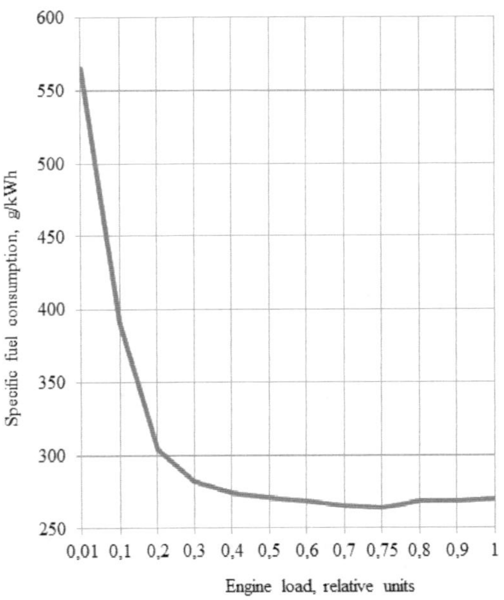

Figure 1. Typical dependence of specific fuel consumption on the diesel generator operating power.

The analysis of the dependences of the specific fuel consumption on the diesel generator operating power shows that the optimum work load of diesel generator units (DGU) of DPPs is provided by maximizing the specific load:

$$P_{DGU}^{sp} = P_{DGU}(t) \Big/ P_{DGU}^{nom} \to \max \tag{17}$$

The condition mentioned above is obviously fulfilled at a specific load closest to 1, i.e. while diesel generators operate with an output power close to the nominal value.

In addition, an important feature of modern DPP control systems is the implementation of an equal specific load distribution:

$$P_{DGU1}^{sp} = ... = P_{DGUn}^{sp} = P_{DPP}^{sp} \tag{18}$$

For unit commitment a matrix of variants is drawn up to provide the maximum operating power of DPP. Table 5 lists an example of unit commitment variants for DPP consisting of 3 diesel generators.

In accordance with (17), one should select a variant with

$$P_{DPP}^{sp} = P_{DPP}^{\max.op} / P_{load}(t) \rightarrow \max$$
$$P_{DPP}^{sp} \leq 1. \tag{19}$$

Additionally, it is recommended to consider the operating hours of each diesel generator engine.

Table 5. Unit Commitment Example for Providing The Maximum Operating Power of The DPP.

Load variant	DGU 1	DGU 2	DGU 3	The maximum operating power of the DPP $P_{DPP}^{\max.op}$
1	1	0	0	P_{DGU1}^{nom}
2	0	1	0	P_{DGU2}^{nom}
3	0	0	1	P_{DGU3}^{nom}
4	1	1	0	$P_{DGU1}^{nom} + P_{DGU2}^{nom}$
5	1	0	1	$P_{DGU1}^{nom} + P_{DGU3}^{nom}$
6	0	1	1	$P_{DGU2}^{nom} + P_{DGU3}^{nom}$
7	1	1	1	$P_{DGU1}^{nom} + P_{DGU2}^{nom} + P_{DGU3}^{nom}$

4. Conclusions

Justification of photovoltaic-diesel hybrid power systems parameters is carried out by modeling operating mode of such systems for various configurations and equipment parameters. The variety of existing approaches to modeling the steady-state operation modes of PV-D system elements leads to the need to analyze their advantages and disadvantages and the possibility of using them in real project activities. In this regard, the article provides an overview and a brief description of simple and easy-to-use mathematical models, the accuracy of which is quite high

5. Acknowledgment

The investigation was carried out within the framework of the project "Development of decision-making software for controlling the operation of a solar-diesel hybrid power system, considering short-term solar radiation forecast" with the support of a grant from NRU "MPEI" for implementation of scientific research programs "Energy", "Electronics, Radio Engineering and IT", and "Industry 4.0, Technologies for Industry and Robotics in 2020-2022".

6. References

[1] Federal Law No 261-FZ 2009 *on Energy Saving and Energy Efficiency Improvement and on Amendments to Certain Legislative Acts of the Russian Federation*

[2] GOST R 55007-2012 *Stationary diesel and gas fueled power plants. Energy efficiency*

[3] Lukutin B V Muravlev I O and Plotnikov I A 2015 *Power supply systems with wind and solar power plants: a work book* Tomsk Publishing House of Tomsk Polytechnic University (in

Russian) [Online] Available https://portal.tpu.ru/SHARED/i/IOM/liter/Tab2/Lukutin_S_elsnab.pdf

[4] *Solar power plants* 2020 Rao-esv.ru. (in Russian) [Online] Available http://www.rao-esv.ru/activity/renewable-energy/solar-power/

[5] Maleki A and Pourfayaz F 2015 *Optimal sizing of autonomous hybrid photovoltaic/wind/battery power system with LPSP technology by using evolutionary algorithms* Sol Energy vol 115 pp 471–483

[6] Smaoui M Abdelkafi A and Krichen L 2015 *Optimal sizing of stand-alone photovoltaic/wind/hydrogen hybrid system supplying a desalination unit* Sol Energy vol 120 pp 263–276

[7] Zondag H 2008 *Flat-plate PV-Thermal collectors and systems: A review* Renew Sustain Energy Rev vol 12 no 4 pp 891–959

[8] Evans D L and Florschuetz L W 1977 *Cost studies on terrestrial photovoltaic power systems with sunlight concentration* Sol Energy vol 19 no 3 pp 255–262

[9] Kant K Shukla A Sharma A and Biwole P H 2016 *Heat transfer studies of photovoltaic panel coupled with phase change materia* Sol Energy vol 140 pp 151–161

[10] Mayer M J and Gróf G 2020 *Extensive comparison of physical models for photovoltaic power forecasting* Appl Energy no 116239 p 116239

[11] Pantic L S Pavlović T M Milosavljević D D Radonjic I S Radovic M K and Sazhko G 2016 *The assessment of different models to predict solar module temperature, output power and efficiency for Nis, Serbia* Energy (Oxf) vol 109 pp 38–48

[12] Hart G W and Raghuraman P 1982 *Simulation of thermal aspects of residential photovoltaic systems* STIN 83

[13] Garg H P and Agarwal R K 1995 *Some aspects of a PV/T collector/forced circulation flat plate solar water heater with solar cells* Energy Convers Manag vol 36 no 2 pp 87–99

[14] Nordmann T Clavadetscher L 2003 *Understanding temperature effects on PV system performance, in Proceedings* of the Third World Conference on Photovoltaic Energy Conversion Osaka Japan pp 2243–2246

[15] Mattei M Notton G Cristofari C Muselli M and Poggi P 2006 *Calculation of the polycrystalline PV module temperature using a simple method of energy balance* Renew Energy vol 31 no 4 pp 553–567

[16] Skoplaki E Boudouvis A G and Palyvos J A 2008 *A simple correlation for the operating temperature of photovoltaic modules of arbitrary mounting* Sol Energy Mater Sol Cells vol 92 no 11 pp 1393–1402

[17] Koehl M Heck M Wiesmeier S and Wirth J 2011 *Modeling of the nominal operating cell temperature based on outdoor weathering* Sol Energy Mater Sol Cells vol 95 no 7 pp 1638–1646

[18] Kratochvil J A Boyson W E and King D L 2004 *Photovoltaic array performance model* Office of Scientific and Technical Information (OSTI)

[19] Akhsassi M et al 2018 *Experimental investigation and modeling of the thermal behavior of a solar PV module* Sol Energy Mater Sol Cells vol 180 pp 271–279

[20] Savkin K D 2020 *Combined autonomous power supply systems using renewable energy sources* MS Thesis Tomsk Polytechnic University [Online] Available http://earchive.tpu.ru/bitstream/11683/38411/1/TPU33084.pdf

[21] Gashimov A M Rakhmanov N R and Akhmedova S T 2013 *Hybrid distributed generation systems with renewable sources: modeling and analysis of their modes of operation in the power system* (in Russian) Izvestiya vysshih uchebnyh zavedenij i energeticheskih ob"edinenij SNG – Energetika [Online] Available https://rep.bntu.by/bitstream/handle/data/5449/%d0%a1.%2021%20-%2031.pdf?sequence=1&isAllowed=y

[22] Stern V I 1980 *Operation of diesel power plants* Moscow Energiya (in Russian)

Method of calculation of water-diverting structures of low-head hydroelectric power plant for power supply of small power consumers

Y Y Zakharov[1], **A R Lepeshkin**[1,2]

[1]National Research University «MPEI», 14 Krasnokazarmennaya, Moscow 111250, Russia
[2]Moscow Aviation Institute (National Research University), 4 Volokolamskoe Shosse, Moscow 125993, Russia

E-mail: lepeshkin.ar@gmail.com, zaharov.uu@mail.ru

Abstract. In recent years and in many countries the economic development of distant regions is increasingly dependent on energy resources. This fact makes the world scientific community pay more attention to the renewable energy sources. Special attention is paid to the solar, wind and small hydropower for electrical consumers who have no possibility to connect to the central power supply lines. In the countries that have water resources the financial support is given to the development of small and micro hydropower stations. The present work presents the results of the research on the improved method of calculation of water-diverting structures of low-head hydroelectric power plant with an installed cross-jet hydro turbine that is actual for the power supply of small power consumers. The presented method can be used for the preliminary analysis of morphometric characteristics of water course as well as the basic parameters of a cross-jet hydro turbine.

1. Introduction

Small hydro power plants are renewable sources of energy, which can be effectively used for the power supply of small power consumers. [1-2].

Micro hydro power plants (MicroHPPs) are one of the first types of HPP in the history of hydraulic power engineering. They were the prototypes of big hydro power plants and often function as models of big hydro turbines. With the active development of the hydraulic power engineering, its main technical decisions started to transmit to small hydraulic power engineering. At the present in the Russian Federation (RF) and many foreign countries the hydro power plants with the capacity of less than 100 kW belong to the category of Micro HPPs [3], while the capacity of one hydraulic unit usually does not exceed 50 kW. The distinction in 100 kW between small and micro hydro power plants is conditional.

In respect to the different natural conditions, it is possible to specify two types of micro HPPs which can realise the potential and kinetic energy of a water course [4]. The examples of the first type are MicroHPPs with classical equipment, river power plants, diversion power plants, hose power plants (a variation of diversion power plants). Micro HPPs of the second type are installed directly in a water course. The examples of these HPPs are the garland HPPs constructed by Blinov B.S. and others, the triplex vertical HPP by Novikov Y.M., rod plane parallel as well as plane lifting by Loginov. M.I. and

Novikov Y.M., membrane, rotary type and capsule hydraulic units which are used abroad.
The science is developing and harland HPPs are pushed into the background. Today the world scientific community sees the perspectives of development of low-head MicroHPPs and is involved in the questions of designing of these systems [5-7], but the given technical solutions are more suitable for plain water flows. New technologies on the gravitational swirl of the water flow [8-10] are advanced but studied not enough. They have some disadvantages such as low efficiency of the hydro turbine in comparison with the cross-flow hydro turbine [11], the necessity to construct a forwarding basin with height difference. These circumstances put some restrictions to the usage of low-head Micro HPPs of the said construction. That is why it is necessary to pay attention to the diversion low-head MicroHPPs which are especially effective for mountainous areas. The present research shows an improved method of calculation of water-diverting constructions of a low-head MicroHPPs, which helps to define the morphometrical characteristics of the water course and also the basic parameters of the cross jet hydro turbine.

2. Method of calculation of water-diverting constructions of a low-head micro HPP

For the effective usage of the water course energy the diversion water lines must provide that the necessary amounts of water pass through with the less possible losses for the operation of the MicroHPP. Diversion water lines can be constructed using open-flow or pressure derivation. The water lines made with open-flow concrete flumes and pressure steel pipes are highly effective. Open-flow flumes (figure 1) are placed depending on the local topography: whether aground whether on the support structures.

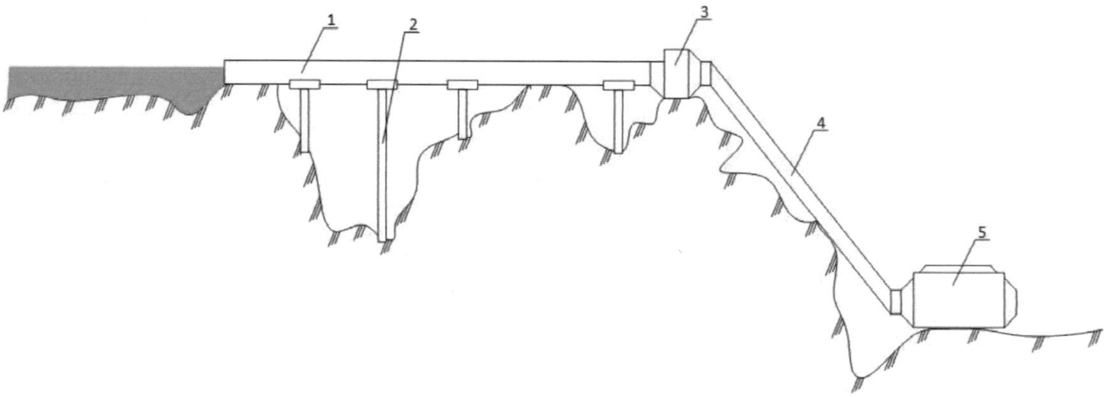

Figure 1. Derivation MicroHPP: 1 – open-flow part of the water line (flume); 2 – support structure of the waterline; 3 – connecting union; 4 – the part of the waterline under pressure; 5 – hydro turbine of the Micro HPP.

Such constructions don't amount to much earthwork which is difficult to perform in a rocky mountainous area. The water lines are made beforehand on a production line basis. The part of the water line under pressure (usually steel pipeline) is installed directly at the descents in front of the MicroHPP, it serves to create the necessary pressure at the hydro turbine. These parts have a relatively small length.
At the construction of the open-flow part the choice of the inclination for the bottom of the line is crucially important as the speed of the water flow depends on it. At small speed of the water flow the water line can silt up or be filled with mud, and in wintertime slush ice and ice sheets can form and jams can occur.
As a rule, silting, slush ice and ice sheets don't occur at the speed of water as:

$$\vartheta_{\text{бу}} \geq |t^{0,06}| \ (\text{м/с}), \tag{1}$$

where t is the calculated minimum ambient air temperature, °C.
On the other hand, at the high speed of the water there are significant head losses and, as a result, the loss of power of the MicroHPP. That is why the speed of the watercourse should be from 1.0 to 1.5

m/s.

The required hydraulic slope of the free-flowing part, m/m, can be calculated by the following formula:

$$i = \frac{\vartheta_{\text{бу}}^2}{C^2 \times r}, \tag{2}$$

where r is a hydraulic radius of the cross section of the free-flowing channel, m; C is a Chezy's velocity factor.

The hydraulic radius of the cross section of the free-flowing channel, m, can be calculated by the formula:

$$r = \frac{S_{\text{бу}}}{\chi}, \tag{3}$$

where $S_{\text{бу}}$ is the cross-sectional area of the flow of the free-flowing part, m^2;

χ – wetted perimeter, m.

The cross-sectional area, m^2, of the flow of the free-flowing part can be calculated by the formula:

$$S_{\text{бу}} = \frac{Q_{\text{ср}}}{\vartheta_{\text{бу}}}, \tag{4}$$

where $Q_{\text{ср}}$ – average annual water flow rate, m^3/s.

The length of the wetted surface side (figure 2), м, is calculated by the formula:

$$a = \sqrt{S_{\text{бу}}}. \tag{5}$$

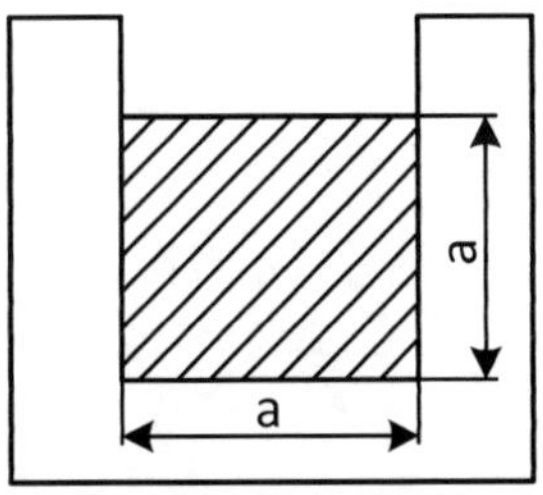

Figure 2. Sectional view of the free-flowing part of the water line.

The wetted perimeter, m, is calculated by the formula:

$$\chi = 3 \times a. \tag{6}$$

In all the calculations of deviation channels of MicroHPPs, the Chezy's velocity factor is calculated according to the formula of N.N.Pavlovskiy:

$$C = \frac{1}{m} \times r^{1/6}, \tag{7}$$

where m is the roughness factor of concrete flumes.

Head loss, m, at the free-flow part of the water line is calculated by the formula:

$$\Delta h = i \times l, \tag{8}$$

l – the length of the free-flow part, m.

Then at the entrance to the pressure part the water head, m, is:

$$H = h_{\text{н}} - \Delta h, \tag{9}$$

where $h_{\text{н}}$ is the initial velocity head, m.

IOP Conf. Series: Materials Science and Engineering 1211 (2022) 012012 doi:10.1088/1757-899X/1211/1/012012

The inner diameter of the pressure pipe line, м, is calculated by the formula:

$$d_{\text{нт}} = \sqrt{\frac{4 \times S_{\text{бу}}}{\pi}}, \tag{10}$$

where π is a mathematical constant.
The average velocity, m/s, of flow in the pressure pipeline is calculated by the formula:

$$\vartheta_{\text{ну}} = \frac{4 \times Q_{\text{ср}}}{\pi \times d_{\text{нт}}^2}. \tag{11}$$

The length of the pressure part, m, is calculated by the formula:

$$L = \frac{H}{cos\alpha}, \tag{12}$$

where α is the angle of the slope of the pressure tunnel in degrees.
The losses of the heads at the pressure area of the tunnel are calculated by the Darcy-Weisbach formula:

$$\Delta H = \lambda \times \frac{L}{d_{\text{нт}}} \times \frac{\vartheta_{\text{ну}}^2}{2 \times g}, \tag{13}$$

where λ –is a resistance factor of the circular pipe;
g – gravity acceleration, m/s^2.
The resistance factor of the circular pipe can be calculated by the formula:

$$\lambda = 8 \times g \times j^2 \times \left(\frac{4}{d_{\text{нт}}}\right)^{3 \times \sqrt{j}}, \tag{14}$$

where j is the roughness factor of the iron pipeline
Then at the entrance to the cross-jet hydro turbine the head of the water, m, will be:

$$H_{\text{т}} = H - \Delta H. \tag{15}$$

The outer diameter of the rotor of the cross jet hydro turbine, m, can be calculated by the formula:

$$d = \sqrt{\frac{Q_{\text{ср}}}{\sqrt{H_{\text{т}}}}}. \tag{16}$$

The width of the rotor, m, is calculated by the formula:

$$b = \frac{Q_{\text{ср}}}{0{,}42 \times d \times \sqrt{H_{\text{т}}}}. \tag{17}$$

Let's calculate the width, m, of the guide blade of the cross-jet hydro turbine:

$$c = 0{,}8 \times b. \tag{18}$$

Let's calculate the height, m, of the guide blade of the cross-jet hydro turbine:

$$x = \frac{Q_{\text{ср}}}{0{,}98 \times c \times \sqrt{2 \times g \times H_{\text{т}}}}. \tag{19}$$

Shaft speed of the rotor, rot/min, of the cross-jet hydro turbine is calculated by the formula:

$$n = \frac{30 \times \sqrt{2 \times g \times H_{\text{т}}}}{\pi \times d}. \tag{20}$$

The shaft energy, kW, of the cross-jet hydro turbine is calculated by the formula:

$$N = 9{,}81 \times Q_{\text{cp}} \times H_{\text{T}} \times \eta, \tag{21}$$

where η is the efficiency coefficient of the cross-jet hydro turbine, %.
Shaft torque, N-m, of the cross-jet hydro turbine is calculated by the formula:

$$M = \frac{9550 \times N}{n}. \tag{22}$$

3. Results

The present improved method of calculation of water-diverting structures of low head micro hydroelectric power plant was developed with the usage of such computer programs as Matlab Simulink and Microsoft Excel. The length of the free-flow part is 30 metres, the angle of the slope of the pressure tunnel is 60 degrees are used as the input data. After entering the data, the dimensions of the free-flow and pressure area of diversion, head of water at the entrance to the cross-jet hydro turbine, main parameters of the rotor of the hydro turbine, capacity, rate of rotation and the shaft torque of the cross-jet hydro turbine were calculated. Figure 3 demonstrates the dependence of the cross-jet hydro turbine capacity on the water flow at different parameters of the initial velocity head. Figure 4 demonstrates the changes of the rotation of the hydro turbine which depend on the water flow at different parameters of the initial velocity head. As it can be seen from the obtained results the capacity of the hydro turbine is directly proportional to the initial velocity head, the effective usage starts at 2.5÷3.0 metres of the initial velocity head. Due to the peculiarities of the construction of the cross- jet hydro turbine the rate of rotation of the rotor remains at the fixed level at different water flows, this fact gives a number of advantages at the year-round usage of MicroHPPs.

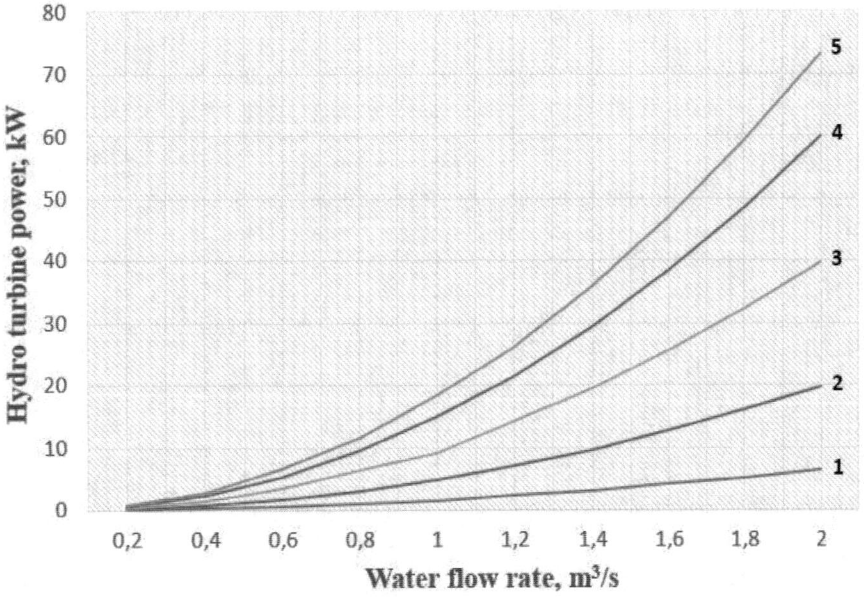

Figure 3. Dependence of the cross-jet hydro turbine power on the water flow at different parameters of the initial velocity head: 1 – 1 m, 2 – 3 m, 3 – 6 m, 4 – 9 m, 5 – 11 m.

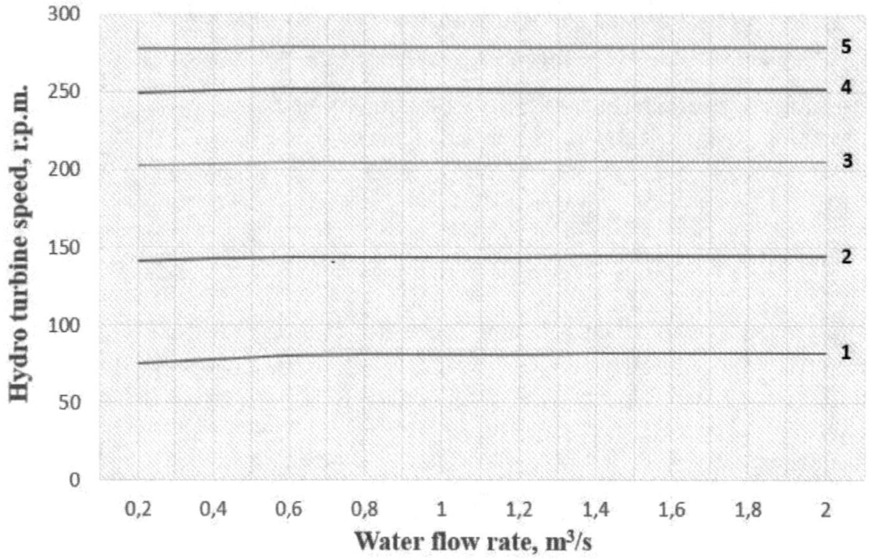

Figure 4. Changes of the rotation of the hydro turbine which depend on the water flow at different parameters of the initial velocity head: 1 – 1 m, 2 – 3 m, 3 – 6 m, 4 – 9 m, 5 – 11 m.

4. Conclusion

Hydro energetical potential for the usage of the MicroHPPs in the RF is not studied good enough. However, the researches that have been carried out in the scientific and technical literature showed that technical capabilities of the energetical usage of the slopes of numerous small rivers (lowland rivers at to 1.7, and mountainous rivers up to 2 thousand kW) can be used by Micro HPPs. The absence of some methodological information at the present time does not allow to give any recommendations to the application of these or that technological decisions at the construction of the Micro HPPs. The present work presents the results of the research on the improved method of calculation of water-diverting structures of low-head hydroelectric power plant with an installed cross-jet hydro turbine that is actual for the power supply of small power consumers. The presented method can be used for the preliminary analysis of morphometric characteristics of water course as well as the basic parameters of a cross-jet hydro turbine.

The necessity in MicroHPP is great and the conditions of its usage vary a lot – from the objects situated at high altitudes in the mountainous region of ex-Soviet Central Asia to small lowlands river lines in the severe conditions of Yakutia.

References

[1] Paish O 2002 Micro-hydropower: Status and Prospects. *Proceedings of the Institution of Mechanical Engineers. Part A: Journal of Power and Energy* **216** 1 31-40 doi.org/10.1243/095765002760024827

[2] Lepeshkin A R and Kachanov A N 2011 Application of Different Constructions of MicroHPPs to Increase the Effectiveness of Electro Supply of Small Power Consumers. *Energo- and Resource Saving – XXI century.: Digest of the VIII International Scientifical and Practical Conf.* (Orel: GTU, Russia) 54-57 https://elibrary.ru/item.asp?id=29177480&pff=1

[3] Mohibullah M, Radzi A M and Hakim M I 2004 Basic Design Aspects of Micro Hydro Power Plant and its Potential Development in Malaysia. *Proc. National Power and Energy Conf. IEEE* 220-223 doi.org/10.1109/PECON.2004.1461647

[4] Mikhailov L P, Feldman B N, Markov T K et al. 1989 *Small Hydropower* (Moscow: Energoatomizdat) p 184

[5] Zhou D and Deng Z D 2017 Ultra-Low-Head Hydroelectric Technology: A review. *Renewable and Sustainable Energy Reviews* **78** 23-30 doi.org/10.1016/j.rser.2017.04.086

[6] Kodirov D and Tursunov O 2019 Calculation of Water Wheel Design Parameters for Micro Hydroelectric Power Station, E3S. *Web of Conf.* **97** 05042 doi.org/10.1051/e3sconf/2019970 5042

[7] Nasir B A 2014 Design Considerations of Micro-Hydro-Electric Power Plant. *Energy Procedia* **50** 19–29 doi: 10.1016/j.egypro.2014.06.003

[8] Yaakob O B, Ahmed Yasser M, Elbatran A H and Shabara H M 2014 A Review on Micro Hydro Gravitational Vortex Power and Turbine Systems. *Jurnal Teknologi* **69** **(7)** 1-7 doi: 10.11113/jt. v69.3259

[9] Sritram P, Suntivarakorn R 2019 The Effects of Blade Number and Turbine Baffle Plates on the Efficiency of Free Vortex Water Turbines. *IOP Conf. Series: Earth and Environmental Science* 012040 257 doi:10.1088/1755-1315/257/1/012040

[10] Rahman M M, Tan J H, Fadzlita M T and Wan Khairul Muzammil A R 2017 A Review on the Development of Gravitational Water Vortex Power Plant as Alternative Renewable Energy Resources. *IOP Conf. Series: Materials Science and Engineering* 012007 217 doi:10.1088/1757-899X/217/1/012007

[11] Adhikari R and Wood D 2018 The Design of High Efficiency Crossflow Hydro Turbines: A Review and Extension. *Energies* **11(2)** 267 doi.org/10.3390/en11020267

Ensuring effective transit of electrical energy using the laws of environmental cybernetics and the foundations of automatic control

A N Kachanov[1], V A Chernyshov[1], A Yu Burova[2], R P Belikov[3], M Sh Garifullin[4]
[1] Oryol State University, 95 Komsomolskaya st., Orel, Russia
[2] Moscow Aviation Institute, 4 Volokolamskoe highway, Moscow, Russia
[3] Orel State Agrarian University, 69 General Rodina st., Orel, Russia
[4] Kazan State Power Engineering University, 51 Krasnoselskaya st., Orel, Russia

Annotation. The necessity of constant monitoring and maintenance of the environmental component in the organization of electric energy transit is substantiated. An original technique aimed at improving the efficiency of the functioning of the air electric network, based on the laws of environmental cybernetics and the basics of automatic control, is considered. The proposed scientific tools allow you to automatically monitor and control all processes that affect the efficiency of electric energy transit from the point of view of its environmental friendliness, and also allows you to form optimal regulatory influences aimed at stabilizing negative processes and disturbing factors arising inside the artificial electrobiological organism "Electric Line - Nature - Man", which violate the stability of its ecological state.

1. Introduction. In turn, the large-scale growth of overhead power lines has a very adverse effect on naturally formed natural landscapes, often leading to the displacement of wildlife from their natural habitat and the suppression of the growth of tree and bush vegetation [1]. Unfortunately, the reality is that the growth of the length and branching of overhead power lines, accompanied by a multiple increase in the amount of energy transmitted, is not always ensured by timely and complete repair and restoration work [2]. As a result, the dynamics of various emergencies on overhead power lines is increasing, which is especially pronounced in recent years, due to the increasing incidence of weather and climatic anomalies [3]. Emergency situations on overhead lines, more and more often become the causes of forest and landscape fires and often lead to fatal electrical injuries and death of animals. Taking into account the fact that the full impact of the growing power grid infrastructure on the natural environment is unlikely to be avoided, the main task of the modern electric power industry is to prevent a point of no return, when the irreversible consequences of this inevitable impact will become one of the causes of the global environmental crisis [4-6]. In this regard, the entire spectrum of negative influence of overhead power transmission lines must be limited in every possible way, as well as its consequences must be calculated and minimized, so that wildlife has all the possibilities and strengths to adapt to them [7]. The principle of the described dynamic homeostasis, in this case, can be briefly formulated as follows – effective provision of environmental sustainability of the artificial electro-biological system "Electrical line - Nature - Man" can be carried out with full control and effective management of all processes (social, biological, technical and other), reflecting the interaction of its elements.

2. Purpose of the study. The main objectives of this research work are approbation and substantiation of the appropriateness of the application of a new, original methodology aimed at ensuring effective transit of electrical energy by maintaining a sustainable ecological balance within the artificial electro-biological organism "Electrical lines - Nature - Man", based on the laws of ecological cybernetics and the basics of automatic process control.

3. Conditions, materials and methods. At the present stage of the formation of scientific and technological progress, characterized by the active introduction of various nature-like technologies in various scientific areas and branches of the industrial sector, research in the field of cybernetics is actively disseminated and popularized as a science that studies the management processes of complex closed systems, primarily with objects of biological origin. One of the characteristic and specific branches of cybernetics, engaged in the scientific study of new methods, forms and means of control in ecology, allowing them to be used in the most effective way, is ecological cybernetics. All the resources of modern ecological cybernetics, at present, are aimed at studying, analyzing and synthesizing complex systems and structures (biotechnical, electrobiological and others), which today do not have a reliable scientific interpretation, or the lack of the possibility of obtaining it. To date, a large number of scientific research works, both domestic and foreign authors, devoted to ecological cybernetics are known, which consider the issues of improving the theory of the system approach and methods of mathematical and physical modeling [8-11].

However, none of the well-known, domestic and foreign works allows to fully, reliably reflect and adequately interpret the patterns and processes occurring inside the artificial electrobiological system "Electrical line-Nature-Man", which makes research in this scientific direction very relevant and in demand. The rapid development of scientific and technological progress, in particular, the development of telecommunication technologies and the evolution of information-computing systems, opens broad horizons before the scientific community for the development of fundamentally new, more effective ways and methods of control of artificial electro-biological and electrical structures. This circumstance allows to minimize the risk of global ecological crisis caused by anthropogenic human activities related to transmission of electric energy through overhead power lines [12-15].

4. Results and discussion. A graphic display revealing the specifics of the functioning of the developed new approach is presented in Figure 1 in the form of a functional automation diagram. The key objects of management of the electrobiological system "Electrical line - Nature - Man" are the following main elements: electrical grid infrastructure (L); sanitary protection zone of the overhead electrical grid (S); plant representatives (F); representatives of the animal world (G); human community (H). The current state and nature of the interaction of elements is controlled by an automatic control system (ACS). It includes the following main components: - meter-recorders (sensors) that generate hardware information from outputs 1-13, which can be stationary (ground), mobile (airborne), including space-based; - information converters SY 1-14; - technological controllers №1 and technological controllers №2; - regulators XC 1-22; - starting stations and actuators NS 1-22 with outputs 14-35. The control information characterizing the dependence of the accident rate on overhead power lines on meteorological characteristics comes from the outputs of the meters-recorders 1-9 through the corresponding information converters SY 1-9 to the process controller No. 1 with the specialized software "Sizam" pre-installed on it [13].

The functioning of the specialized software "Sizam" is based on the foundations of mathematical modeling, based on statistics on accidents at overhead electrical lines and data characterizing weather and climatic factors over the past five years. Specialized software allows you to quickly and reliably determine the number of failures per hundred kilometers of the overhead electrical grid, characterized by different voltage, length and technical condition. At the same time, the algorithm for determining the number of failures is different for the warm and cold seasons.

The authors selected the following factors as the most significant weather and climatic factors that significantly affect the degree of influence of an element of the Electrical grid infrastructure (L) on the sustainability of the ecological state of the artificial electro-biological system "Electrical Line - Nature - Man" [15]:

For the warm season, valid from the beginning of April to the second decade of October, inclusive: X1 - average surface temperature (°C); X2 - average wind speed (m/s); X3 is the amount of precipitation (mm); X4 - duration of thunderstorm activity (h).

For the warm season, valid from the beginning of the third decade of October to March inclusively: X'1 - average air temperature (°C); X'2 - average wind speed (m/s); X'3 - amount of precipitation (mm); X'4 - number of days with ice load (days);

The regression equation characterizing the dependence of the number of incidents associated with the failure of overhead electrical lines and the violation of the stability of the functioning of the artificial electro-biological system "Electrical line-Nature-Man" (Y) on four weather and climatic factors (X1, X2, X3, X4) is identified by the mathematical equation of the first order [16, 17]:

$$Y = b_0 + b_1X_1 + b_2X_2 + b_3X_3 + b_4X_4 + b_{12}X_1X_2 + b_{13}X_1X_3 + b_{14}X_1X_4 + b_{23}X_2X_3 + b_{24}X_2X_4 + \\ + b_{34}X_3X_4 + b_{123}X_1X_2X_3 + b_{124}X_1X_2X_4 + b_{134}X_1X_3X_4 + b_{234}X_2X_3X_4 + b_{1234}X_1X_2X_3X_4 \quad (1)$$

Previous studies made it possible to determine with high accuracy the coefficients of the equation for various categories of overhead electrical lines, characterized by different lengths and technical conditions, as a result of which the necessary series of corresponding equations was obtained [18].
For example, in the warm season, with an average length of overhead transmission lines and a good technical condition, the number of accidents is determined as follows:

$$Y = 0{,}10174 + (0{,}00424X_1) + (0{,}00180X_2) + (0{,}00428X_4) + \\ (-0{,}00189X_1X_2) + (-0{,}00230X_1X_4) + (0{,}00174X_1X_2X_3X_4) \quad (2)$$

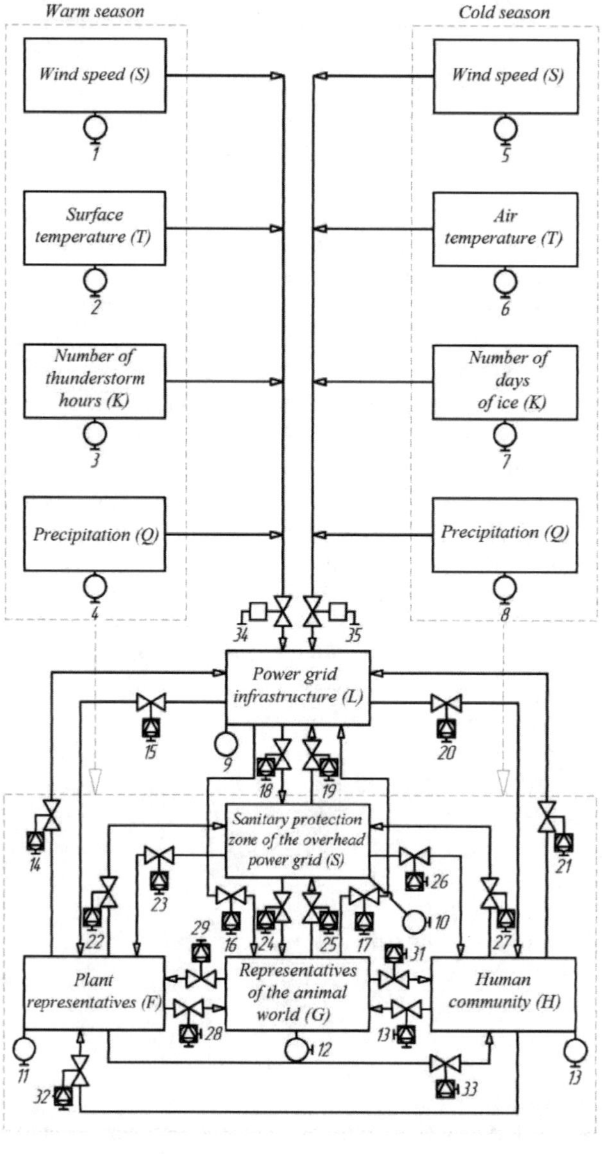

Figure 1. Figure 1. Functional diagram of automatic control of the artificial electro-biological system "Electrical line - Nature - Man ", providing it with a stable ecological state

Continuation of figure 1

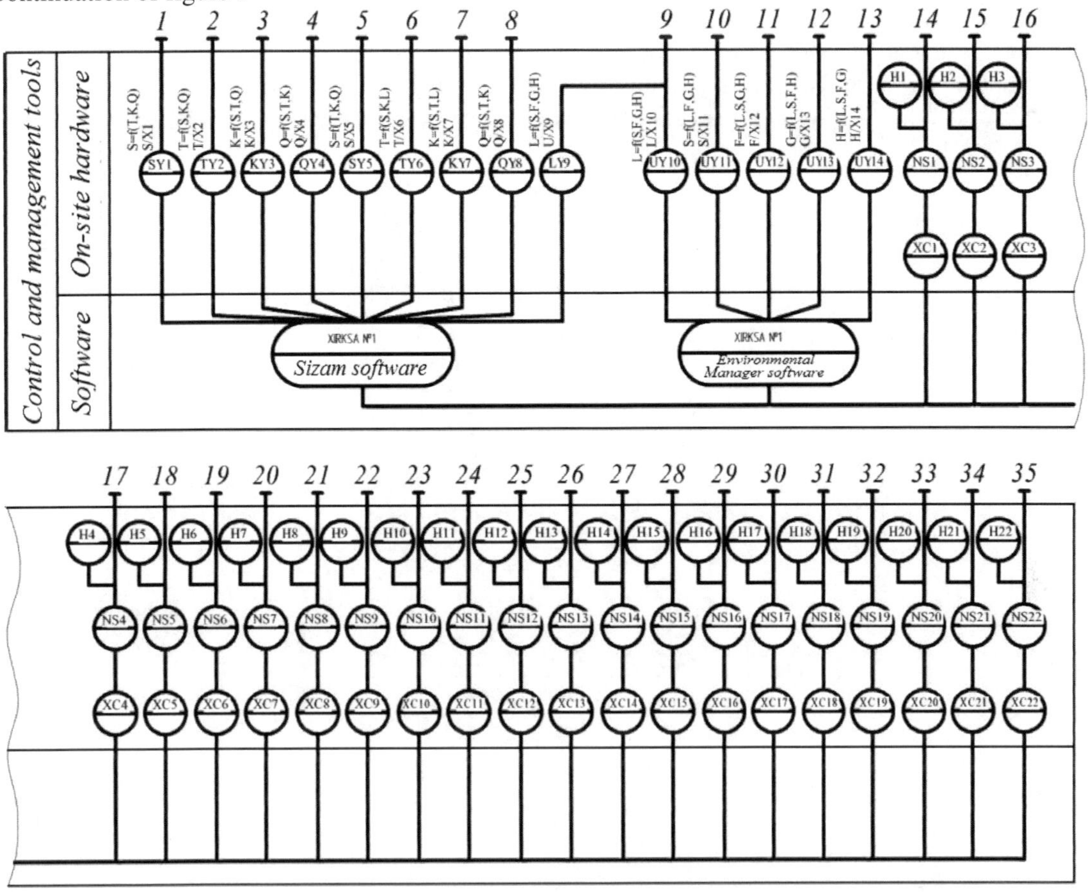

Technological controller technological controllers №1, with pre-installed specialized software "Manager-ecologist" [14], through UY 10-14 converters, receives hardware information that qualitatively and quantitatively characterizes the state of elements (L; S; F; G; H), via various communication channels, coming from measuring recorders with outputs 9-13.

Fulfilling the algorithm laid down by the authors, aimed at ensuring a stable ecological state of the artificial electro-biological system "Electrical line - Nature - Man", the technological controller generates commands to the relevant regulatory bodies, launching means and actuators to stabilize negative processes and disturbing factors that arise within this system. It should be noted that the functioning of the specialized software "Manager-Ecologist" is based on the classical, cybernetic foundations of flow graphs, which allows qualitatively and quantitatively to make a reliable assessment of the influence of the electrobiological system on each element, as well as the influence of the elements on each other and on the entire system as a whole. At the same time, both direct influence and indirect (through other elements) are assessed [19-20].

The block of flow graphs presented on the functional diagram clearly displays the degree of interaction of five elements: overhead electrical grid infrastructure (L); sanitary protection zone of the overhead electrical grid (S); plant representatives (F); representatives of the animal world (G); human community (H) [12].

Analysis of the generated flow graph shows that it contains ten cycles formed by two vertices (Θ2.1-Θ2.10), twenty cycles formed by three vertices (Θ3.1-3.20), thirty cycles formed by four peaks (4.1-4.30), as well as 24 cycles formed by five tops (5.1-5.24). In this case, the following effects of the influence of individual cycles are obtained: Θ2.1-Θ2.10, Θ3.1-Θ3.20, Θ4.1-Θ4.30, Θ5.1-Θ5.24.

The combined effect of these cycles can be expressed as [21]:

$$\Theta = 1 / (1- (\Theta2.1\text{-}\Theta2.10) - (\Theta3.1\text{-}\Theta3.20) - (\Theta4.1\text{-}\Theta4.30) - (\Theta5.1\text{-}\Theta5.24) \quad (3)$$

Thus, for the considered artificial complexly organized electro-biological system, it is possible to determine the degree of influence of individual elements on each other.

For example, the overhead electrical grid infrastructure (L) can be affected by the elements: the human community (H), the sanitary protection zone of the overhead electrical grid (S); plant representatives (F); representatives of the animal world (G).

$$L \leftarrow H = \left[\frac{H1 + H2S1 + H3G1 + H4F1}{1 - \sum_{i=1}^{10} \Theta2_i - \sum_{i=1}^{20} \Theta3_i - \sum_{i=1}^{30} \Theta4_i - \sum_{i=1}^{24} \Theta5_i} \right] \quad (4)$$

$$L \leftarrow S = \left[\frac{S1 + S2H1 + S3G1 + S4F1}{1 - \sum_{i=1}^{10} \Theta2_i - \sum_{i=1}^{20} \Theta3_i - \sum_{i=1}^{30} \Theta4_i - \sum_{i=1}^{24} \Theta5_i} \right] \quad (5)$$

$$L \leftarrow F = \left[\frac{F1 + F2S1 + F3H1 + F4G1}{1 - \sum_{i=1}^{10} \Theta2_i - \sum_{i=1}^{20} \Theta3_i - \sum_{i=1}^{30} \Theta4_i - \sum_{i=1}^{24} \Theta5_i} \right] \quad (6)$$

$$L \leftarrow G = \left[\frac{G1 + G2S1 + G3H1 + G4F1}{1 - \sum_{i=1}^{10} \Theta2_i - \sum_{i=1}^{20} \Theta3_i - \sum_{i=1}^{30} \Theta4_i - \sum_{i=1}^{24} \Theta5_i} \right] \quad (7)$$

The complex effect of an artificial electro-biological system on the overhead electrical grid infrastructure (L) is determined by the formula:

$$L \leftarrow HSFG = \left[\frac{S1(1 + H2 + F2 + G2) + H1(1 + S2 + F3 + G3) + F1(1 + H4 + S4 + G4) + G1(1 + H3 + S3 + F4)}{1 - \sum_{i=1}^{10} \Theta2_i - \sum_{i=1}^{20} \Theta3_i - \sum_{i=1}^{30} \Theta4_i - \sum_{i=1}^{24} \Theta5_i} \right] \quad (8)$$

The influence of the human community (H) on the artificial electro-biological system "Electrical line-Nature-Man" is determined by the expression:

$$LSFG \leftarrow H = \left[\frac{H1 + H2 + H3 + H4}{1 - \sum_{i=1}^{10} \Theta2_i - \sum_{i=1}^{20} \Theta3_i - \sum_{i=1}^{30} \Theta4_i - \sum_{i=1}^{24} \Theta5_i} \right] \quad (9)$$

Considering the fact that it is not realistic to obtain the degree of influence on the electro-biological system of the parameters F1-F4, G1-G4, H1-H4, S1-S4, L1-L4, even through continuous and widespread observation, since this will require colossal material and technical costs, the authors consider it expedient to use linguistic variables, qualitatively and quantitatively reflecting the degree of influence of any quantity, first used in the scientific work of J. Ostergaard [22].

Table 1 summarizes the membership functions for the proposed use of linguistic concepts.

Membership functions are set for continuous values of variables; in practice, you can use their point estimates: mean values, variances, medians, etc.

Table 2 summarizes the values of modes, mathematical expectations and standard deviations corresponding to the membership functions presented in table 1.

The use of the input evaluation information presented in Table 2 will provide the control automation with the ability to effectively assess the stability of the functioning of the electro-biological system "Electrical line-Man-Nature" directly on the basis of linguistic terms.

At the same time, the technological controllers №2 technological controller will form the appropriate proactive commands for the XC 1-22 regulatory bodies (structural divisions and services), to activate the necessary NS 1-22 launch resources and actuators with outputs 14-35, implementing effective measures of continuous, intermittent and relay control actions, which ensure the stabilization of the ecological balance of the electrobiological system "Electrical line-Man-Nature".
As executive mechanisms used to stabilize the ecological balance, various organizational, technical and other measures can be used to ensure the protection of flora and fauna, the safety of human life, and also increase the operational reliability of the overhead electrical transmission line.

Table 1. The membership functions of the proposed linguistic concepts

Linguistic term	Membership Function Expressions	Linguistic term	Membership Function Expressions
"Very positive"	$1-\exp\left[-\left(\frac{0,25}{\lvert 1-x\rvert}\right)^{2,5}\right]$	"Neutral"	$1-\exp\left[-\left(\frac{0,25}{\lvert x\rvert}\right)^{2,5}\right]$
"Positive"	$1-\exp\left[-\left(\frac{0,25}{\lvert 0,75-x\rvert}\right)^{2,5}\right]$	"At least negative"	$1-\exp\left[-\left(\frac{0,25}{\lvert -0,1+x\rvert}\right)^{2,5}\right]$, $если 0<x<1$
"Fairly positive"	$1-\exp\left[-\left(\frac{0,25}{\lvert 0,5-x\rvert}\right)^{2,5}\right]$	"Seemingly negative"	$1-\exp\left[-\left(\frac{0,25}{\lvert -0,25-x\rvert}\right)^{2,5}\right]$
"Seemingly positive"	$1-\exp\left[-\left(\frac{0,25}{\lvert 0,25-x\rvert}\right)^{2,5}\right]$	"Fairly negative"	$1-\exp\left[-\left(\frac{0,25}{\lvert -0,5-x\rvert}\right)^{2,5}\right]$
"At least positive"	$1-\exp\left[-\left(\frac{0,25}{\lvert 0,1-x\rvert}\right)^{2,5}\right]$, $если 1<x<0$ $1-x, если 0<x<1$ $1+x, если -1<x<1$	"Negative"	$1-\exp\left[-\left(\frac{0,25}{\lvert -0,75-x\rvert}\right)^{2,5}\right]$
		"Very negative"	$1-\exp\left[-\left(\frac{0,25}{\lvert -1-x\rvert}\right)^{2,5}\right]$

Table 2. Values of modes, mathematical expectations and standard deviations, corresponding to membership functions

"Linguistic term"	E(x)	G(x)	Мода	"Linguistic term"	E(x)	G(x)	Мода
"Very positive"	0,64	0,24	1,0	"Neutral"	0,00	0,44	0,00
"Positive"	0,55	0,30	0,75	"At least negative"	-0,13	0,44	-0,1
"Fairly positive"	0,44	0,35	0,5	"Seemingly negative"	-0,23	0,37	-0,25
"Seemingly positive"	0,23	0,37	0,25	"Fairly negative"	-0,44	0,35	-0,5
"At least positive"	0,13	0,37	0,1	"Negative"	-0,55	0,3	-0,75
				"Very negative"	-0,64	0,24	-1,0

If an incorrect or ineffective work is detected, any of the activated executive measures, a complete prohibition, partial blocking or correction of further work will be introduced, for the corresponding ineffective mechanism, through managerial orders H 1-22, formed by operators (authorized experts), manually carrying out continuous selective control over the ecological state of the electro-biological system "Electrical line-Nature-Man", which, if necessary, can also control and implement the functions of information converters SY 1-14. According to the canons of classical automation, automatic control systems (ACS) are capable of providing the following three control options: 1 - by the deviation of the controlled value; 2 - by disturbance of the controlled value; 3 - combined [23]. The process of interaction of the elements of the artificial electro-biological system "Electrical transmission line-Nature-Man", one way or another, proceeds with the appearance of various kinds of disturbing factors characterized by positive, neutral or negative influences, including weather and climatic ones, which can entail a violation of the ecological equilibrium of the electrobiological system. The developed criterion for the operational safety of overhead electrical transmission lines and the criterion for the environmental safety of overhead electrical transmission lines [24, 25] allow to control the efficiency of functioning of the artificial electrobiological system "Electrical line-Man-Nature". Figure 2, a, shows the ACS operating according to the disturbance deviation algorithm, the work of which is that if, under the action of external disturbances, the excess (deviation) between the actual and prescribed criterion (value) of the controlled variable will exceed the limit set by the operator, then the input of the regulating device will receive such a reference action at which this deviation will decrease.

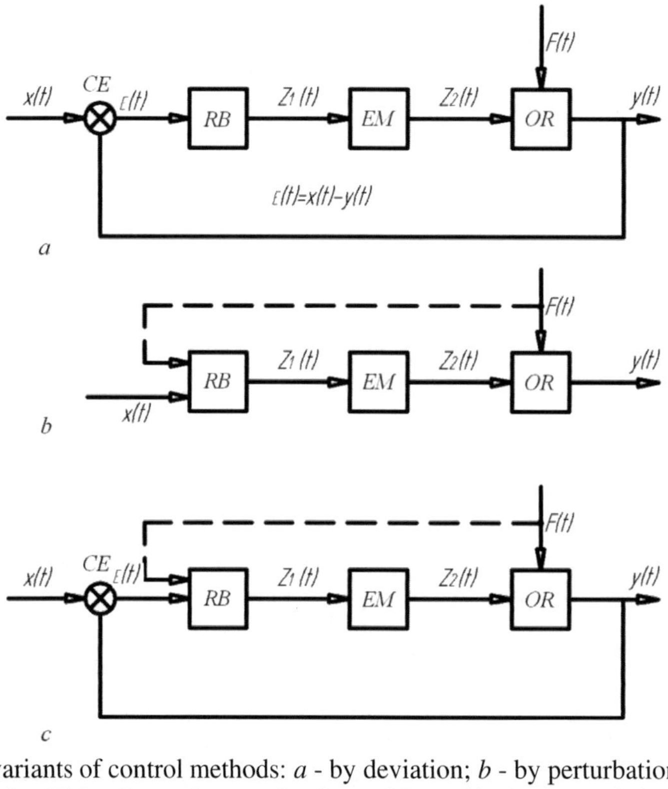

Figure 2. Possible variants of control methods: *a* - by deviation; *b* - by perturbation; *c* – combined. *RB* - regulatory body, *EM* - Executive mechanism, *OR* - object of regulation, *CE* - comparison element, x (t) - setting action, Z1 (t) and Z2 (t) - internal control actions, y (t) - controlled value, F (t) - disturbing action.

Any of the executive mechanisms (measures) aimed at ensuring a stable ecological state of the artificial electro-biological system " Electrical line - Nature - Man" is activated when the controlled value is deviated, and this action is directed in such a way that it always reduces the deviation.
The ES comparison element is introduced in order to obtain the difference in values [23]:

$$\varepsilon\,(t) = x\,(t) - y\,(f) \quad (1)$$

The perturbation control algorithm (perturbation compensation) is based on the principle of compensating for the influence of a change in the perturbing effect by using a special device.
Figure 2, b shows the ACS operating according to the disturbance compensation algorithm.

Taking into account the fact that there can be several disturbing influences and they can be initiated by different reasons, in practice it is advisable to limit ourselves to considering the disturbing influences initiated only by the main cause, since otherwise the operation of the automatic control system becomes significantly more complicated.

Figure 2, c shows the ACS operating according to the combined control algorithm, which combines the two previously considered algorithms, respectively, according to the deviation of the disturbing value and according to the disturbance. The combined regulation algorithm is widely used, as a rule, in the construction of more serious ACS that implement high quality regulation, for example, in critical situations, when the harmonious existence of its elements is threatened within the framework of the functioning of the “Electrical line-Nature-Human” electro-biological system. Based on Figure 2, it follows that for any method of regulation, the automatic control system consists of two parts, the regulated part is the object of regulation and the regulating part is the hardware part.

The regulating, hardware part includes the following elements: meter-recorder, which controls the current value of the controlled variable; a comparison element that determines the amount of deviation of the controlled value from the prescribed value; regulating organ, responsible for formation of corresponding signals, providing stabilization of regulated value; actuating mechanism, responsible for setting regulated value to the previous level, after its deviation.

During continuous control of the nature of interaction of the elements of the artificial electro-biological system "Electric Line - Nature - Man", by means of ground, air and space based measuring-recorders, the control hardware signals, through various communication channels, arrive at the input of the controlling device continuously. However, the control actions arriving at the input of regulators and providing them with a necessary algorithm of functioning can be continuous or discontinuous. The curve of deviation Δu, the controlled variable y, in time from the set value y0, is shown in Fig. 3, and the lower part of the figure shows how the control action Z for the actuator activated in operation is changing continuously.

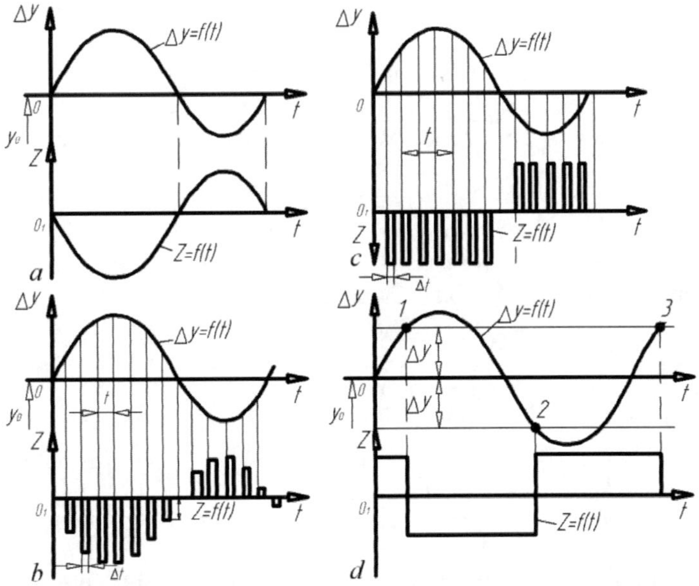

Figure 3. Main regulating actions and their schemes: a - continuous, b, c - discontinuous, d - relay

The regulating action, in the first version, is displayed by a series of pulses of the same duration Δt, following at regular intervals t1 = t2 = t, while the magnitude of the pulses Z = f (t) is proportional to the value of the control signal at the time of the formation of the control action. The regulating action, in the second version, is displayed by pulses of the same magnitude Z = f (t), following at the same time intervals t1 = t2 = t, but they have different duration Δt, which is determined by the value of the control signal at the moment of the formation of the control impact. In this case, the regulatory action from the regulatory body to the executive mechanism comes with appropriate interruptions, as a result of which the executive mechanism forms a regulatory action also intermittently. Also find relay control systems (see Figure 3, d), the regulatory bodies of which ensure the principle of two-position regulation, characterized by two stable positions. The relay controller switches to the first position, when the deviation of the regulated quantity exceeds the set positive limit + Δy, and to the second position it switches when the deviation changes the sign and reaches the set negative limit - Δy. At the same time, it should be noted that in each situation the control actions are formed, which are the same in absolute value, but differ from each other in sign. On the basis of the above, it follows that control using a relay-regulator causes sharp activation of the controlled object, in connection with the fact that the absolute value of the deviation, in the process of functioning of the controlled object, always decreases. In case the deviation Δu changes and reaches the acceptable positive value + Δu (point 1), the relay regulator generates a signal through which a corresponding regulating influence Z, equal in value to the positive value of the regulating influence + Z, but differing from it by a sign, is delivered to the controlled object through the regulating organ and the executive mechanism. In the preset time interval the regulated quantity deviation will decrease, while the deviation value Δu (point 2) will become equal to the permissible negative value -Δu and the technological controller will issue a corresponding control signal, and the regulating action Z will change its sign to the opposite one accordingly, and so on. If we compare relay control with other control methods, its circuitry is simple to implement and cheap, so it is advisable to use it where there are no serious requirements for regulation, i.e. when high sensitivity is not needed to respond to disturbing influences.

5. Conclusions. Based on the presented justification of expediency of cybernetic automatic control system, with a detailed consideration of the specifics of functioning of its theoretical and circuit parts, it follows that it allows to provide reliable and continuous control of the environmental situation and the formation of effective regulatory impacts aimed at stabilizing it. In this regard, we can, with a high degree of probability, believe that the engineering and technical solution presented by the authors, aimed at ensuring the effective transit of electric power through overhead power grids of different voltage classes, using the principles of environmental cybernetics and automatic control, can interest advanced power supply companies, having the desire and resources to minimize anthropogenic impact initiated by their infrastructure, as well as increase the operational reliability of overhead lines.

Reference list

[1] Kachanov A.N., Chernyshov V.A., Meshkov B.N., Garifullin M.Sh., Pechagin E.A. Improving the efficiency of 6-10 kV electrical networks with isolated neutral in conditions of single-phase earth faults. Sustainable Energy 2021 International Symposium (SUSE-2021), Kazan, Russia February 18-20, 2021 https://doi.org/10.1051/e3sconf/202128801010

[2] Vinogradov A.V., Tikhomirov D.A., Bolshev V.E., Vinogradova A.V., Sorokin N.S, Borodin M.V., Chernyshov V.A., Golikov I.O., Boukreev A.V. Features of distributed electrical engineering integration in agriculture. Intelligent Computing and Optimization. ICO 2020. Advances in intelligent systems and computing, vol. 1324. Springer, Cham. In kn .: Vasant P., Zelinka I., Weber G.W. (ред.) https://doi.org/10.1007/978-3-030-68154-8_2

[3] Kachanov A.N., Chernyshov V.A., Revkov A.A., Borodin M.V., Vladimirov O.V. Prediction of single-phase faults in 10 kV electrical grids with regard to seasonal influence of weather factors. Sustainable Energy Systems: Innovative Prospects. SES 2020. Lecture Notes in Civil Engineering. vol. 141, pp. 310-320, 2021. https://doi.org/10.1007 / 978-3-030-67654-4_34

[4] Environmental aspects of electric electrical transmission. V. I. Chekhov / Edited by G. K. Zarudsky. - Moscow: MPEI Publishing House, 1991. - 44 c.

[5] Fedorov V.D. Changes in natural biological systems / Edited and commented by Professor V.N. Maximov. - Moscow.: Publishing house "RAGS", 2004. - 368 c

[6] Bolshakov V.N. Ecological forecasting / V.N. Bolshakov. - Moscow: Znanie, 1983. - 62 c. : ill. - (New in life, science, technology) (Biology ; No 5). - Bibliography: p. 60

[7] Traulko E.V. Ecological bases of nature management and ecology of health: tutorial.
Traulko E.V. Ecological bases of nature management and ecology of health: textbook. - Novosibirsk: Novosibirsk State Technical University Press, 2017. - 196 c.
[8] Hayes D. Causal analysis in statistical research / D. Hayes - M .: Finance and Statistics., 1981. - 225 c.
[9] Eliseeva I.I. Logic of applied statistical analysis / I.I. Eliseeva, V.O. Rukavishnikov. - Moscow: Finance and Statistics, 1982. - 193 c.
[10] Taganov I. N. Causal analysis of complex systems / I. N. Taganov, G. A. Turgumbaev. - Almaty: Mektep, 1984. - 127 c.
[11] Zakharova Y.A. Methods for assessing the complex safety of installations, workshops and enterprises with incomplete information / Y.A. Zakharova, V. Shubin // Safety of Work in Industry. - 1996 - №8. - C. 39-43.
[12] Patent № 2676889 Russian Federation, IPK G06G 7/62. Meteomonitoring system for forecasting the probability of damage to electric grid equipment and evaluation of preventive and recovery works / Yu.I. Magaraz, A.Yu. Khostantsev, S.A. Panfilov; applicant and patent holder PJSC "Moscow Unified Electric Grid Company" (PJSC "MOESK"). - No. 2018100741; application. 10.01.2018; published 11.01.2019, Bulletin No. 2.
[13] Certificate of state registration of computer program № 2008610196 RF. Calculation of the expected number of earth faults in 10 kV distribution networks with isolated neutral SIZAM (synoptic) / Chernyshov V.A., Gavrichenko A.I.; co-applicant Orel State Agrarian University. ¹ 200761427; declared on 29.10.2007; publ. 09.01.2008.
[14] Certificate of state registration of computer programs No. 2011616873 RF. "ECOLOGY-EXPERTS (for system Electrical line - Human - Nature) -ecological examination of the biotechnical system of Electrical transmission lines-Man-Nature" / Chernyshov VA, Chernyshova LA; applicant Oryol State Agrarian University. No. 2011615036; app. 07.07.2011; publ. 09/06/2011.
[15] Mizintseva M.F. Information ecology / M.F. Mizintseva, L.M. Koroleva, V.V. Bondar // Specialized meteorological support of branches of the economy: teaching materials for the course "Economic meteorology" for students studying in the field of 05.03.05 Applied hydrometeorology / S.N. Lapina, M.A. Alimpieva. - Saratov: Publishing Center "Nauka", 2020. - 50 c.
[16] Rumshinsky Ya.Z. Mathematical processing of experimental results / Ya.Z. Rumshinsky. - Moscow: Nauka, 1971. - 192 c.
[17] Sigorsky V.P. Mathematical apparatus of an engineer / V.P. Sigorsky - Kiev: Technika., 1975. - 768 c.
[18] Kachanov A.N. Theoretical aspects of short-term prediction of single-phase earth faults in overhead distribution networks of 10 kV. Kachanov, V.A. Chernyshov, E.A. Pechagin, A. Kobelev // News of higher educational institutions of Chernozem region 2020 ¹2 (60). C. 28-38.
[19] Wiener N. Cybernetics, or Control and Communication in Animals and Machines / 2nd edition - Moscow: Soviet Radio Publishing House, 1968. - 326 c.
[20] Melnikov V.N. Ecological Cybernetics. Fundamentals of Control of Ecological Processes and Systems. Melnikov V.N., Melnikov A.V. // Astrakhan: Astrakhan State Technical University, 2010. - 384 c.
[21] Korshunov Yu.M. Mathematical Foundations of Cybernetics: Textbook for Higher Education Institutions. - Third edition, revised and enlarged. - Moscow: Energoatomizdat, 1987. - 496 c.
[22] Ostergaard J / - J. Fuzzy logic control of heat exchanger processes. In Gupta M., Saridis G. and Gaines B / (Eds), Fuzzy Automata and Decision Processes, North-Holland Publishing Co.
[23] Fundamentals of automatic regulation and control. Ed. Ponomarev V.M. and Litvinov A.P. Educational. guide for non-electrical. specialties of higher education institutions. Moscow: Vysshaya shkola Publisher, 1974 - 439 p.
[24] Chernyshov V.A. Justification of criteria of anthropogenic and environmental safety of overhead electrical lines / V.A. Chernyshov, L.A. Chernyshova // Agrotechnics and Electrical supply. Orel, 2014. no. 2 (2). C. 60-71.

Modeling the system of a dovetailless asynchronous electric drive for conveyor mechanisms

Vladimir Vladimirovich Pikalov, Andrei Igorevich Boikov, Andrei Vitalievich Sdvizkov, Pavel Sergeevich Ponomarev

Lipetsk State Technical University Automation and computer science faculty

Abstract. The article considers a vector control system for asynchronous electric drive of belt conveyors, which uses an adaptive observer to control the speed of the electric drive. This allows you to increase the reliability and inter-interval maintenance of the electric drive. The requirements for the observer are determined, its mathematical description is made, and the law of adaptation is determined. In the Matlab Simulink package, a computer simulation of the proposed sensor-free control system is performed. A study of the robustness of the actuator in terms of the drift of the parameters of the motor when powered from the inverter, expressed by the integral criterion for parametric robustness. Based on the results obtained, a comparison is made and conclusions are drawn about the advantages and disadvantages of using an adaptive observer in comparison with other types of observers.
Keywords. Electric drive, induction motor, inverter, vector control system, adaptive observer, robustness, Matlab Simulink package.

1. Introduction

Currently, one of the main requirements for mechanisms is system reliability and an increase in the period of trouble-free operation. Often, without losing the quality of the mechanism, it becomes almost impossible to achieve an increase, but there is a possibility when the modernization of the mechanism can lead to an increase in the reliability of the system and an increase in interval maintenance. One such method is to abandon the scalar control system with a speed sensor on the shaft in favor of a vector control system, without installing a speed or position sensor. It is proposed to consider the application of such a control system on the PB-1 belt conveyor installed at the PJSC Mikhailovsky GOK. The use of such a system will make it possible to abandon the speed sensor and reduce the electromechanical vibrations that occur during the conveyor operation. This paper will consider a belt-type conveyor, which is used at the enterprise of PJSC "Mikhailovsky GOK" in the workshop for the production of high-quality concentrate and shipment of finished products in the building of the concentrate warehouse. Conveyor PB-1 is designed to transport iron ore concentrate to a warehouse. During the operation of the belt conveyor, due to the elasticity of the belt, vibrations of the transported material occur, which negatively affects the condition of the bearings on the entire conveyor system and the service life of this unit as a whole. To combat this phenomenon, it is proposed to use a vector control system, which includes correction factors. The vector control system itself can be used both with feedback from a speed sensor mounted on the shaft and without installing a sensor by establishing feedback from an observer. The speed sensor also has its resource and requires maintenance, and in case of its failure, the operation of the entire conveyor stops indefinitely until the malfunction is eliminated. To eliminate this drawback and increase reliability, it is proposed to use a vector sensorless drive control system.

In the course of this work, the types of observers will be considered, criteria for their choice are compiled and comparative modeling is performed in the MATLAB Simulink software shell, and the robustness of each of the selected observers is investigated.

2. Observers' requirements

To compile a list of observers' understudies, the minimum requirements were formed that must be met by control systems with an observer:

- maximum accuracy of state variable recovery;
- minimum sensitivity to the deviation of model parameters;

- no cumbersome mathematical calculations;
- accessibility in understanding the approach, the possibility of its physical interpretation.

According to these criteria, it can be concluded that the observer should be with an adaptive controller to adapt to thermal changes in parameters and not consume large computing power. Analyzing available sources by criteria, adaptive observers were selected for the research:

- rotor flux linkage observer;
- EMF observer;
- stator current observer;
- full order observer.

The selected observers are adaptive, that is, they have an adaptation unit for changing parameters, therefore, during the operation of the drive when it heats up, there will be no "collapse" of the observer's work, and the value of the static error between the value received from the observer with the value received from the sensor will not abruptly increase during drive operation. Also, for the work of the selected observers, there is no need for large computing power, which increases the speed of calculations and the response of the observer.

3. Materials and methods

The frequency converter can be taken as an aperiodic link with a constant time having a dependence on the PWM frequency. In most cases, the used frequency converters have a modulation frequency fPWM=2 kHz. This frequency will be taken for the further calculation of the control system. Then

$$T_d = \frac{1}{f_{pwm}}.$$

Calculating the gain of a transistor inverter:

$$k_d = \frac{k_s \cdot \sqrt{2} \cdot U_{fn}}{U_n},$$

where $k_s = 1,5$ – UF frequency converter safety factor,

$U_{fn} = U_{lin.n}/\sqrt{3}\ V$ – rated phase voltage of the motor,

$U_n = 5\ V$ – maximum control circuit reference voltage.

Thus,

$$W_{UF}(p) = \frac{k_d}{(T_d p + 1)}.$$

Calculating the stator current vector

$$\bar{i}_s = I_{sn} \cdot cos\varphi + jI_{sn} \cdot sin\varphi.$$

No-load current vector calculation:

$$\bar{i}_0 = j\frac{U_{fn}}{\omega \cdot L_s}.$$

Rotor flux is presented as

$$\overline{\Psi_r} = \frac{1}{k_r} \cdot \overline{\Psi_0} - L_{r\sigma} \cdot \bar{i}_s,$$

where $\overline{\Psi_0} = \bar{i}_0 \cdot L_m$ – vector of the main flux linkage of the motor. The maximum value of the rotor flux linkage modulus.

$$\Psi_{rm} = \sqrt{2} \cdot |\Psi_r|.$$

UΨ flux link sensor transmission ratio

$$k_\Psi = \frac{U_{\Psi m}}{\Psi_{rm}}.$$

UA current sensor transfer ratio

$$k_c = \frac{U_{mm}}{I_{sm}} = \frac{U_{mm}}{\lambda \cdot \sqrt{2} \cdot I_{sn}}.$$

The transfer function of the open-loop regulation of the magnetizing component of the stator current has the form

$$W_{TSd}(p) = W_{AAd}(p) \cdot \frac{k_d}{(T_d p + 1)} \cdot \frac{(1/R_s') \cdot k_c}{(T_s p + 1)},$$

where $W_{AAd}(p)$ – current regulator transfer function.

The AAd regulator parameters are calculated and adjusted according to the modular optimum conditions. PI controller is selected with a transfer function as a current regulator

$$W_{AAd}(p) = \frac{\tau_t p + 1}{T_{it} p}.$$

As a result, getting the following values

$$\tau_t = T_s = \frac{L_s'}{R_s'} = \frac{\sigma \cdot L_s}{R_s + k_r^2 \cdot R_r};$$

$$T_{it} = \frac{2 \cdot T_d \cdot k_d \cdot k_t}{R_s'}.$$

Since the parameters of the internal circuits of an induction motor do not depend on the projections of the fixed axes, the structure and parameters of the AAq current regulator are identical to AAd parameters.

When optimizing the control loop of the rotor flux linkage module, only one large time constant $T_r = L_r / R_r$ taking into account, since $T_{s1} > T_{s2}$ and $T_{s2} \cong T_{s0}$. This allows obtaining the transfer function of the open-loop control

$$W_{0TSd}(p) = W_{A\Psi}(p) \cdot \frac{\left(\frac{1}{k_c}\right) \cdot \varepsilon \cdot L_m \cdot k_\Psi}{(2T_d p + 1) \cdot (T_{s1} p + 1)},$$

where $W_{A\Psi}(p)$ – transfer function of the rotor flux linkage modulus regulator.

The coefficient which is characterizing the ratio of the reduced to the rotor and the actual stator resistance is calculated as

$$\varepsilon = R_s' / R_s.$$

A Ψ regulator parameters are calculated and set according to the modular optimum. PI controller with a transfer function as the regulator of the flux linkage module is selected

$$W_{A\Psi}(p) = \frac{\tau_\Psi p + 1}{T_{i\Psi} p} = k_{d\Psi} + \frac{1}{T_{i\Psi} p}.$$

As a result of optimization, the following values are obtained

$$\tau_\Psi = T_r = \frac{L_r}{R_r};$$

$$T_{i\Psi} = \frac{4 \cdot T_d \cdot \varepsilon \cdot L_m \cdot k_\Psi}{k_t};$$

$$k_{d\Psi} = \frac{\tau_\Psi}{T_{i\Psi}}.$$

The open-loop transfer function of motor speed control is shown as

$$W_s(p) = W_{AR}(p) \cdot \frac{\left(\frac{1}{k_c}\right)}{(2T_d p + 1)} \cdot \frac{3 \cdot p_d \cdot k_c \cdot k_s \cdot \Psi_r}{2 \cdot Jp},$$

where $k_s = (U_{cm} \cdot p_d)/\omega_0$ – transmission ratio of BR tachogenerator,

ω_0 / p_d – the idle speed of the motor at the maximum frequency of the UF inverter.

Optimization of the speed loop can be carried out by including a P-controller (modular optimum) or a PI-controller (symmetric optimum) in the loop. In this case, the transfer function of the P-controller is determined by the expression

$$W_{AR}(p) = k_{sr} = \frac{J \cdot k_c}{6 \cdot k_r \cdot T_d \cdot k_s \cdot p_d \cdot \Psi_r}.$$

The transfer functions of the VF vector filter with the obtained values will have the following form

$$W(p)_1 = \frac{\sigma \cdot L_s \cdot p_d}{k_d \cdot k_c \cdot k_s} \cdot (T_d p + 1);$$

$$W(p)_2 = \frac{k_r \cdot p_d}{k_d \cdot k_\Psi \cdot k_s} \cdot (T_d p + 1).$$

4. Model reference adaptive control (MRAC) observer

In the model reference adaptive control (MRAC) observer, the motor speed is observed using a reference model and an adjustable model. The operating principle of the observer is shown in Figure 1. The reference model, which is independent of motor speed, calculates the state variable using the measured voltages and currents. An adaptive model that depends on the motor speed calculates the state variable $\tilde{x}$. Then the difference (ε) between the state variables is used to control an adaptive mechanism that restores speed $\tilde{w}_{el}$.

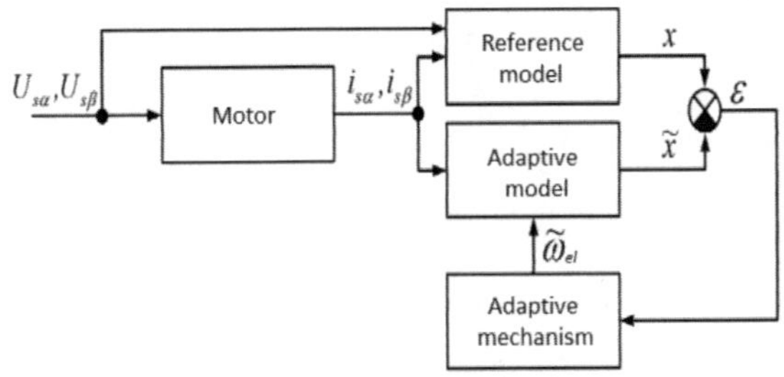

Figure 1. The principle of organizing the observer based on the adaptive model (MRAC).

Nowadays, there are many methods for observing motor speed. One such method is model reference adaptive control (MRAC) based observer. This method is quite an interesting approach, simple and easy to implement.

Velocity observation methods using the Model Adaptive Observer (MRAC) can be classified according to the different types of state variables used. Among these methods, the most commonly used is the observer, whose variable is the rotor flux linkage. The structural diagram of the observer and its mathematical model (MRAC) are shown in figures 2 and 3.

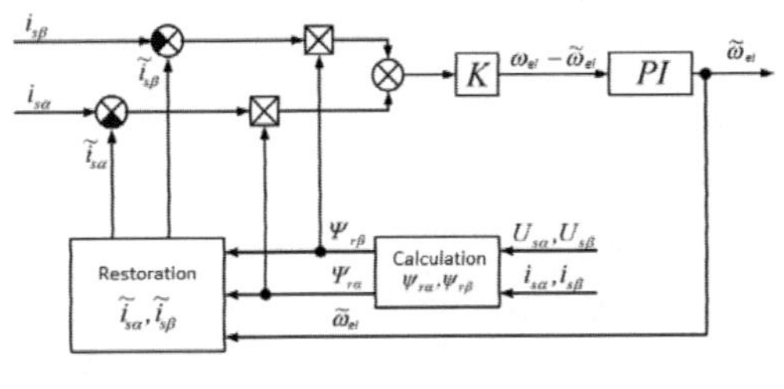

Figure 2. Structural diagram of the observer based on the adaptive model (MRAC).

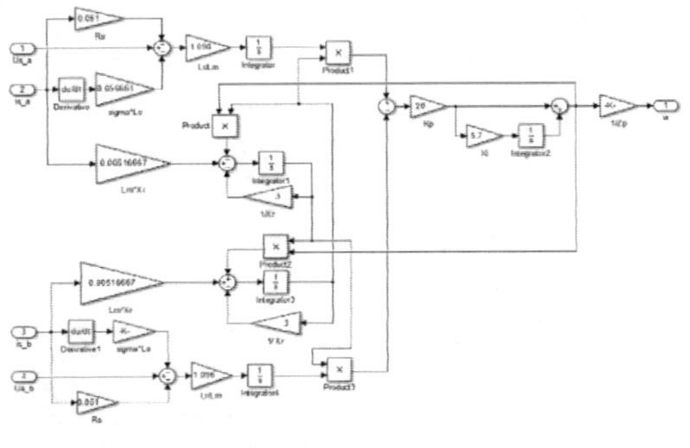

Figure 3. Observer mathematical model based on adaptive model (MRAC).

The mathematical model and simulation results are shown in figures 4, 5, 6, 7.

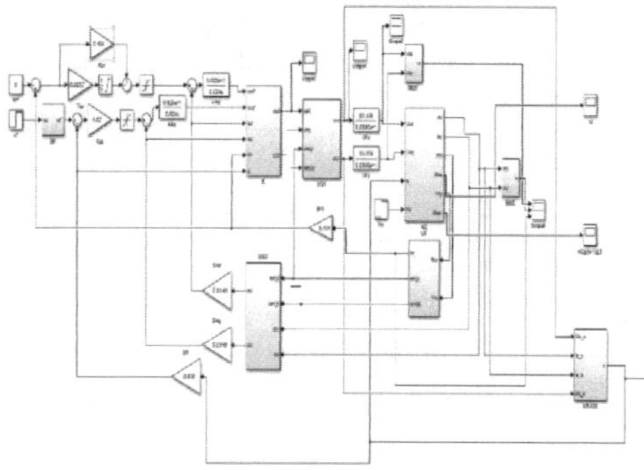

Figure 4. Mathematical model of vector control systems with an observer-based adaptive model (MRAC).

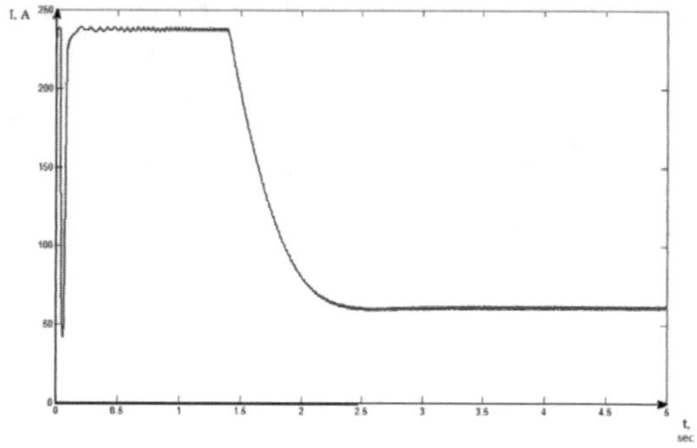

Figure 5. Current characteristic for vector control systems with an observer-based on the adaptive model (MRAC).

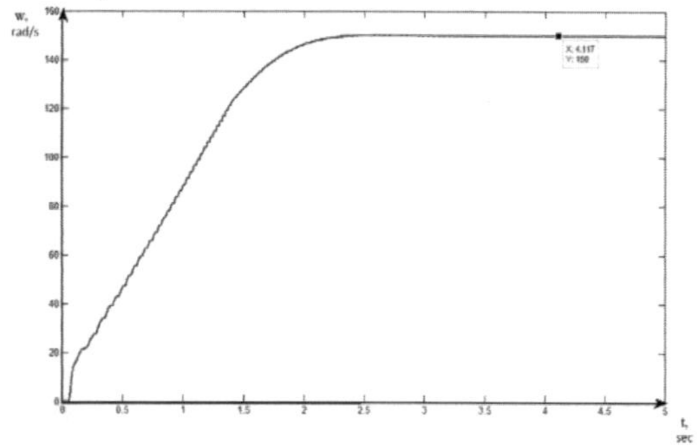

Figure 6. Current characteristic for vector control systems with an observer-based on the adaptive model (MRAC).

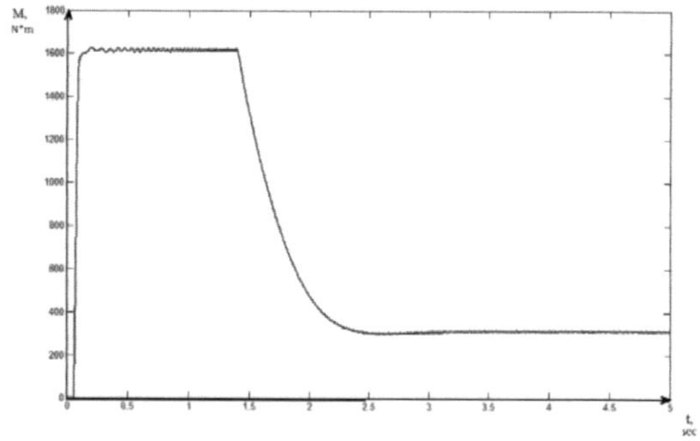

Figure 7. Torque characteristic for vector control systems with an observer-based on an adaptive model (MRAC).

In a real vector control system, it is never possible to accurately set the parameters of the motor, in addition, some of them change during operation, for example, the thermal change in the active resistance of the stator and rotor. Therefore, the main task of modeling was to assess the robustness of the system when changing the motor parameters.

In a real system, during motor operation, a change in active resistance occurs, which negatively affects the quality of the observer's work, and the resulting error leads to the observer falling apart. The change in the rotor resistance does not have a significant effect on the operation of the system, therefore, in the future, the change in the stator will only be considered. The observer's work in normal mode and a reduction in stator resistance by 20% are shown in figures 8, 9.

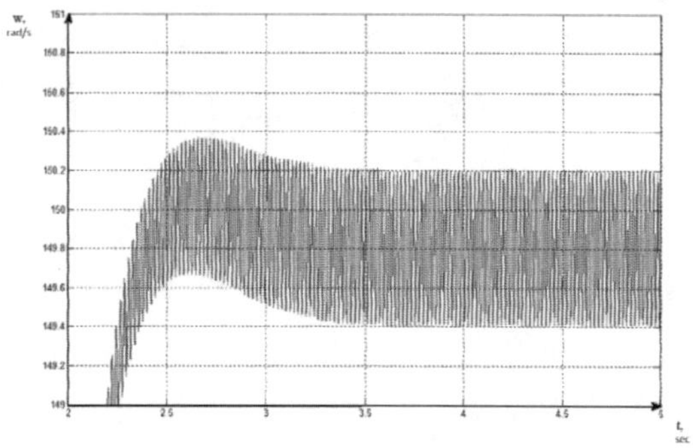

Figure 8. Velocity characteristic for vector control systems with an observer-based adaptive model (MRAC) at reduced stator resistance.

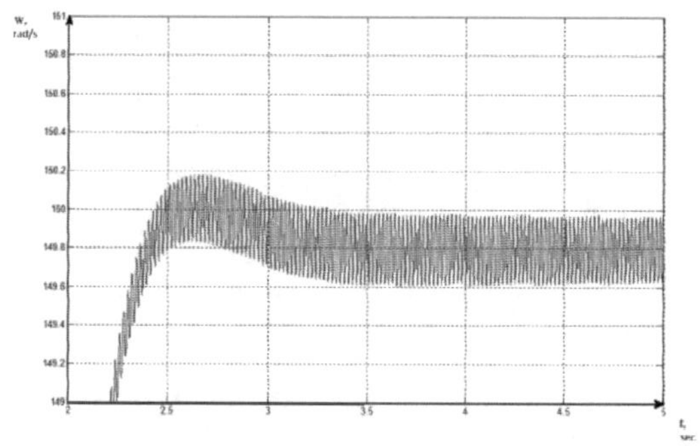

Figure 9. Velocity characteristic for vector control systems with an observer-based adaptive model (MRAC) at reduced resistance.

Also, the disadvantage of this system is the impossibility of operating the system in conditions where the load on the motor is applied at the time of operation by a sharp throw, the way out of this problem may be to start the motor already under load.

As shown by modeling, this method turned out to be quite robust (insensitive) to changes in almost all parameters of the IM, except for the active resistance of the stator. The high sensitivity of the system to Rs is due to the fact that this parameter is included in the integration operation (the integration operation tends to accumulate error) in the equation of the reference model of the given observer. The advantage of this method for use in vector control systems is its simplicity. However, for the operability of such a system, a mechanism (online) for identifying the active resistance of the stator is required.

The methods based on the adaptive tuning of the MRAC motor model are quite simple to understand and easy to implement. At the same time, often when solving differential equations included in the mathematical description of the observer based on the adaptive model (MRAC), the integration operation is used, which creates certain difficulties in calculations and accumulates an error in the operation of the system.

5. Conclusion

In this work, the vector control system of an electric drive is simulated using the example of the PB-1 conveyor using adaptive observers. In the work, observers were considered for the stator current, for the rotor flux linkage, for the EMF, and the observer of full order.

The choice of a control system that meets the compiled criteria was carried out, mathematical modeling of the vector control system was carried out, and a comparative analysis of the systems was made using observers and with a speed sensor on the shaft. Each of the observers was examined for a robustness parameter.

An assumption is made about the applicability of a full-order observer on the PB-1 conveyor as a system that increases the reliability of the unit.

References

[1] V.M. Terekhov, O.I. Osipov, Sistemy upravleniya elektroprivodov, M.: Izdatel'skii tsentr «Akademiya», 2005, p. 304.

[2] G.G. Sokolovskii, Elektroprivody peremennogo toka s chastotnym regulirovaniem, M.: Izdatel'skii tsentr «Akademiya», 2006, p. 272.

[3] Ando Takeshi, Tanamachi Tokunosuke, Toiota Eiichi, Nakata Kiioshi, Suzuki Masato, Yasuda Koudzhi, " Ustroistva ili sposoby upravleniya elektrodvigatelyami upravleniem vektora, naprimer putem upravleniya orientatsiei polya," RU patent 2193814, 27.11.2002.

[4] Afanas'ev Kirill Sergeevich, Glazyrina Tat'yana Anatol'evna, Glazyrin Aleksandr Savel'evich, Kozlova Lyudmila Evgen'evna, Timoshkin Vadim Vladimirovich, "Ustroistvo upravleniya asinkhronnym dvigatelem," RU patent 2438229, 27.12.2011.

[5] Ionov Aleksei Aleksandrovich, Zelenin Dmitrii Vadimovich, Kozachenko Vladimir Filippovich, Alyamkin Dmitrii Ivanovich, Lashkevich Maksim Mikhailovich, "Sposob bezdatchikovoi otsenki uglovogo polozheniya rotora mnogofaznogo elektrodvigatelya," RU patent 2428784, 10.09.2010.

[6] Yu.N Kalachev, Nablyudateli sostoyaniya v vektornom elektroprivode (zapiski diletanta), M., 2015, p. 60.

[7] A.C. Isakov, Realizatsiya nablyudatelya sostoyanii asinkhronnogo dvigatelya s korotkozamknutym rotorom v bezdatchikovoi sisteme vektornogo upravleniya, *Nauchno-tekhnicheskii vestnik. SPbGU ITMO*, 2007, no. 38, pp. 280-286.

[8] L.V. Shelomkova, D.I. Alyamkin, Sistema vektornogo bezdatchikovogo upravleniya asinkhronnym dvigatelem s pereklyuchaemoi strukturoi, *Elektrichestvo*. 2008, no. 5. pp. 30-35.

[9] A.S. Anuchin, Sistemy upravleniya elektroprivodov. M.: Izdatel'skii dom MEI, 2015, p. 373.

[10] A. Vinogradov, A. Sibirtsev, I. Kolodin, Adaptivno-vektornaya sistema upravleniya bezdatchikovogo asinkhronnogo elektroprivoda serii EPV, *Silovaya elektronika*, 2006, no. 3, pp. 50-55.

[11] A.V. Kuksin, A.V. Romanov, Matematicheskaya model' adaptivno-vektornoi sistemy upravleniya bezdatchikovogo asinkhronnogo elektroprivoda, *Elektrotekhnika*, 2009, no. 11. pp. 15-20.

[12]V.N. Meshcheryakov, Dinamika elektromekhanicheskikh sistem pod"emnotransportnykh mekhanizmov s asinkhronnym elektroprivodom, Lipetsk: LGTU, 2002, p. 120.

[13]M.G. Bychkov, "*Fundamentals of theory, control and design of gate-inductor electric drive*" Ph.D. dissertation (in Russian), MPEI, Moscow, Russian Federation, 1999. p. 372.

[14]N. Kaewpoo, K. Ohyama, Y. Nakazawa, H. Fujii, H. Uehara, Y. Hyakutake, "Simulation of SRM Sensorless Control System for Electric Vehicle," in *2018 International Conference on Engineering, Applied Sciences, and Technology (ICEAST)*, Jul. 2018, pp. 1-4, DOI: 10.1109/ICEAST.2018.8434508

[15]C. Pavlitov, H. Chen, I. Colak, T. Tashev, Y. Gorbounov, "Sensorless control of SRM by the aid of artificial neural network adaptive reference model," in *Proceedings of the 2011 14th European Conference on Power Electronics and Applications*, 15 Sept. 2011, pp. 1-7.

[16]M. Rafiee, E. Afjei, A. Siadatan, "Design and construction of a new SRM sensorless driver," *in 2013 21st Iranian Conference on Electrical Engineering (ICEE)*, Feb. 2013, pp 1-5.

[17]L. He, H. Sun, J. Gao, J. Bai, "A new method to detect the initial position for sensorless SRM," in *2011 International Conference on Electronics, Communications and Control (ICECC)*, 9-11 Sept. 2011, pp. 2450-2453, DOI: 10.1109/ICECC.2011.6067879

[18]J. Cai, Z. Deng, "Initial Rotor Position Estimation and Sensorless Control of SRM Based on Coordinate Transformation," in *IEEE Trans. Instrum. Meas.*, Apr. 2015, vol. 64, no. 4, pp. 1004-1018, DOI: 10.1109/TIM.2014.2364699

[19]V.A. Bondarko, Adaptivnoe vektornoe upravlenie asinkhronnym elektrodvigatelem na osnove metoda rekurrentnykh tselevykh neravenstv, *Avtomatika i telemekhanika*, 2010, no. 9. pp. 120–135.

[20]V.V. Pankratov, Vektornoe upravlenie asinkhronnymi elektroprivodami, Novosibirsk: Izd-vo NGTU, 2019, p. 6.

Information about the authors:

Vladimir V. Pikalov, Automation and computer science faculty Lipetsk State Technical University Lipetsk, Russia;
eLibrary SPIN: 2211-6081
E-mail: wreditels@mail.ru

Andrei I. Boikov, Automation and computer science faculty Lipetsk State Technical University Lipetsk, Russia;
eLibrary SPIN: 6011-7793; ORCID: 0000-0002-0032-0683

Andrei V. Sdvizkov, Automation and computer science faculty Lipetsk State Technical University Lipetsk, Russia;
E-mail: andrei_sdvizhkov@mail.ru

Pavel S. Ponomarev, Automation and computer science faculty Lipetsk State Technical University Lipetsk, Russia;
E-mail: pavel.sergeevi4.ponomarev@gmail.com

IOP Conf. Series: Materials Science and Engineering 1211 (2022) 012015 doi:10.1088/1757-899X/1211/1/012015

Low-temperature induction heating of flat metal objects in a traveling electromagnetic field

A N Kachanov, Y S Stepanov, N A Kachanov, V A Chernyshov and D A Korenkov

Russia, Oryol State University named after I.S. Turgenev, Russia, 302026, Komsomolskaya st., 95

E-mail: kan@ostu.ru

Annotation. The article discusses possible options for a low-temperature induction heating system (LTIHS) of flat metal products in a traveling electromagnetic field. The problem of calculating eddy currents, active and reactive powers induced in a heated flat object, as well as electromagnetic forces acting on the object moving it in a given direction, is posed and solved. A mathematical model has been developed that takes into account the dependence of the influence on the main parameters of the electromagnetic field of the following factors: geometric dimensions of the air gap between the poles of the magnetic circuit and the heated flat body; the longitudinal edge effect caused by the open circuit of the magnetic circuit of the inductor, as well as the transverse edge effect associated with the appearance of the longitudinal components of eddy currents in a heated flat object. The solution of particular problems of LTIHS in one- and two-dimensional formulation allows them to be simplified and to perform calculations for various design variants of induction heating devices with a traveling electromagnetic field, using a one-dimensional model that explicitly takes into account the features of electromagnetic processes in the systems under study.

1. Introduction

In various sectors of the economy, when carrying out a number of technological processes, the systems for low-temperature induction heating of flat products in a traveling electromagnetic field can be successfully used, shown in Figure 1 (a, b, c). The installation provides not only heating of the object 2, but also of the media inside the heated object, with their simultaneous movement in space [1,2,3,4].

2. Materials and methods

In the study of LTIHS, the problem of calculating eddy currents, active and reactive powers induced in a heated flat object and electromagnetic forces moving the object in a given direction was posed and solved. In the general case, the solution to the problem posed requires taking into account a number of specific features, such as: change in the parameters of the electromagnetic field when the heated object is removed from the inductor; longitudinal edge effect due to the open circuit of the magnetic circuit of the inductor; transverse edge effect associated with the appearance of longitudinal components of eddy currents in a heated flat object; redistribution of eddy currents due to the geometric dimensions of the heated object and the displacement of induced currents due to the skin effect; the effect of the resulting electromagnetic force on a heated flat object, depending on the method of switching on the inductor windings (counter or concordant switching on).

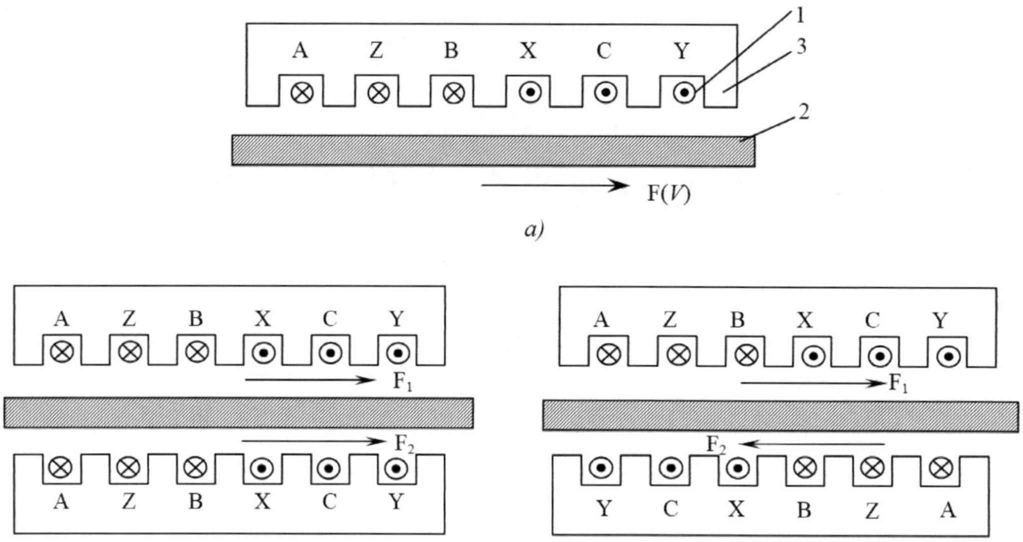

Figure 1. The main types of systems "inductor - flat heated metal object"
1 - inductor; 2 - heated object; 3 - magnetic circuit.

To find a general solution to the problem posed, we will use the superposition method, which allows us to consider a complex three-dimensional problem in one- and two-dimensional formulation. Let us single out the two most characteristic methods of heating flat products in a traveling electromagnetic field - one-sided and two-sided heating. As a basic model, we will take the one-dimensional mathematical model shown in Figure 2.

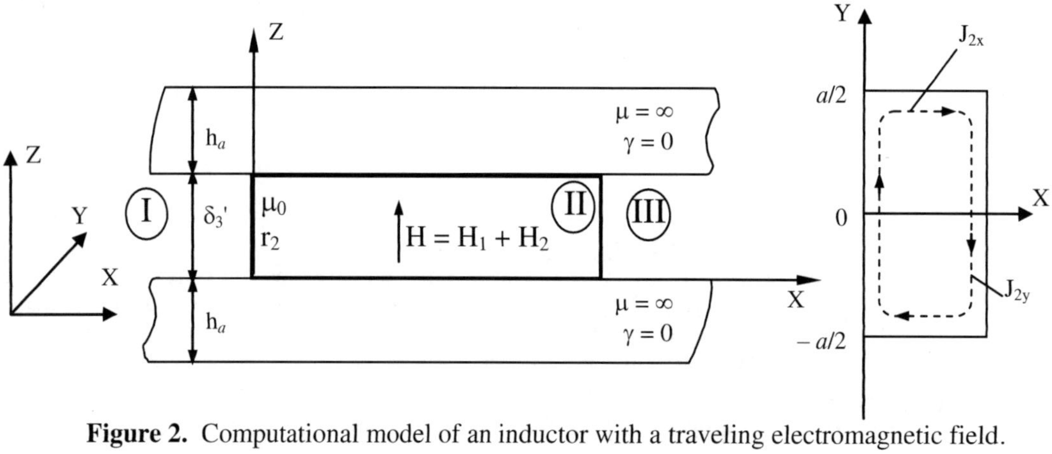

Figure 2. Computational model of an inductor with a traveling electromagnetic field.

The following assumptions are made for the design model:

1. The magnetic cores of the upper and lower inductors have infinite dimensions along the X and Y axes. The parameters of the electromagnetic field do not depend on the Y and Z coordinates.

2. The magnetic permeability of the cores of the inductors $\mu = 0$. The cores of the inductor are laminated, which allows us to take their electrical conductivity in the transverse direction (along the Y-axis) equal to zero ($\gamma_1 = 0$).

3. Primary and secondary currents are evenly distributed along the height of the air gap (δ_3) with densities J_1 and J_2, respectively.

4. The currents of the inductor winding, located in the slots of the magnetic circuit with a height (h_a), form a sinusoidally traveling current wave in the interpolar space, the bulk density of which is equal to [5]:

$$J_1 = J_{1m} \cos(\omega t - \alpha x), \tag{1}$$

or in complex form

$$\dot{J}_1 = J_{1m} e^{-j\alpha x}, \tag{2}$$

where $\alpha = \pi/\tau$ is a wave number;

$$J_{1m} = J_{\phi} \frac{\sqrt{2} m w_{\phi} k_{об}}{p \tau \delta_3^{'}} = J_{nm} / \delta_3^{'}, \tag{3}$$

where $J_{ф}$ – effective value of the phase current of the inductor winding; m – number of phases; $w_{ф}$ – the number of turns of the phase winding of the inductor; $k_{об}$ – winding ratio; p – number of pole pairs; τ – pole division; J_{nm} – the amplitude of the linear current density of the inductor.

5. A real flat metal object of length *b* (along the X-axis), width *a* (along the Y-axis), thickness *d* (along the Z-axis), with magnetic permeability μ_0 and specific electrical conductivity γ_2 is replaced by an equivalent one that completely fills the equivalent working gap of the inductor (δ'_3) on a segment of length *b* and having an equivalent conductivity:

$$\gamma_2^{'} = \gamma_2 \frac{d}{\delta_3^{'}}. \tag{4}$$

6. The active surfaces of the magnetic cores of the inductor are assumed to be smooth without taking into account the grooves.

Taking into account the accepted assumptions, the magnetic field strength in the working gap of the inductor will have only one component along the Z axis, which is equal to $\dot{H} = \dot{H}_z = H_1 + H_2$. The electric field strength and current density J_1 and J_2 have components only along the X-axis. For real induction devices with at least 6 poles, as shown in [6], the influence of the longitudinal edge effect can be neglected.

The change in the strength Hz along the Z coordinate for various design variants of inductors with a traveling magnetic field are taken into account in the design model when determining the equivalent air gap $\delta_3^{'}$. As an equivalence criterion, the equality of the values of the strength of the external (primary) magnetic field at the location of the heated flat object is taken:

$$\dot{H}_1 = \frac{1}{d} \int_{\delta}^{\delta+d} \dot{H}_z(z) dz, \tag{5}$$

where $\dot{H}_z(z)$ - the complex value of the normal component of the primary field strength in the working gap of real LTIHS products in a traveling electromagnetic field (Figure 1 a, b, c), and $\dot{H}_1$ is the complex of the magnetic field strength in the computational model (Figure 2).

To determine $\dot{H}_z(z)$ we will use a two-dimensional computational model proposed by A.I. Voldek [5], in which the complex value of the normal component of the tension for the system under consideration is equal to:
for one-sided heating (Fig. 1, *a*)

$$\dot{H}_z(z) = -jJ_{nm}e^{-j\alpha x}e^{-\alpha z}, \tag{6}$$

for double-sided heating of the plate when the windings of the upper and lower inductors are concordantly switched on (Fig. 1, *b*)

$$\dot{H}_z(z) = -j\frac{J_{nm}}{2}\frac{ch\alpha(z-\delta_3/2)}{sh(\alpha\delta_3/2)}e^{-j\alpha x} \tag{7}$$

for double-sided heating of the plate with the opposite connection of the windings of the upper and lower inductors (Fig. 1, *c*)

$$\dot{H}_z(z) = -jJ_{nm}\frac{ch\alpha(\delta_3 - z)}{sh\alpha\delta_3}e^{-j\alpha x} \tag{8}$$

In the proposed calculation model, the magnetic field strength, which is a constant value along the height of the working gap of the inductor, can be calculated by the formula:

$$\dot{H}_1 = -j\frac{J_{nm}}{\alpha\delta_3'(1+\beta^2/\alpha^2)}e^{j\alpha x}, \tag{9}$$

where β is a coefficient of propagation of the resulting magnetic field in the working gap of the inductor

$$\beta = \sqrt{\mu_0/h_a\delta_3'}. \tag{10}$$

Taking into account the condition of equivalent calculation models (5) and on the basis of (6) - (7), we find the expression for the air gap ($\delta_3 = \delta + d$):
for the case of one-sided heating of the plate

$$\delta_3' = \frac{de^{\alpha\delta}}{(1+\beta^2/\alpha^2)(1-e^{-\alpha d})}; \tag{11}$$

for double-sided heating of the plate with the consent of the upper and lower inductors

$$\delta_3' = \frac{d}{(1+\beta^2/\alpha^2)sh(\alpha d/2)}\left[cth\left(\frac{\alpha\delta_3}{2}\right)ch\frac{\alpha(2\delta+d)}{2} - sh\frac{\alpha(2\delta+d)}{2}\right]; \tag{12}$$

for double-sided heating of the plate with the opposite connection of the upper and lower inductors

$$\delta_3' = \frac{d}{2(1+\beta^2/\alpha^2)sh(\alpha d/2)}\left[cth\alpha\delta_3\,ch\frac{\alpha(2\delta+d)}{2} - sh\frac{\alpha(2\delta+d)}{2}\right]; \tag{13}$$

where δ is the distance from the poles of the magnetic circuit to the surface of the heated flat object in a real system.

3. Results and discussion

The possibility of solving particular problems in one- and two-dimensional formulations made it possible to simplify the task and perform calculations for various variants of induction heating devices with a traveling electromagnetic field, using a one-dimensional model that allows explicitly taking into account the features of electromagnetic processes in the systems under study.

Dividing the core of the model (Figure 2) into three areas, taking into account the accepted assumptions, we write the following system of ordinary second-order linear differential equations with constant coefficients:

$$\left.\begin{array}{l} \text{for zone 1} \\ \dfrac{\partial^2 \dot{A}_{1m}}{\partial x^2} - \gamma_1 \mu_0 v \dfrac{\partial \dot{A}_{1m}}{\partial x} - j\omega\gamma_1\mu_0 \dot{A}_{1m} = 0 \\ \text{for zone 2} \\ \dfrac{\partial^2 \dot{A}_{2m}}{\partial x^2} - \gamma_2 \mu_0 v \dfrac{\partial \dot{A}_{2m}}{\partial x} - j\omega\gamma_2\mu_0 \dot{A}_{2m} = -\mu_0 \dot{J}_{1m} e^{-j\alpha x} \\ \text{for zone 3} \\ \dfrac{\partial^2 \dot{A}_{3m}}{\partial x^2} - \gamma_3 \mu_0 v \dfrac{\partial \dot{A}_{3m}}{\partial x} - j\omega\gamma_3\mu_0 \dot{A}_{3m} = 0 \end{array}\right\} \tag{14}$$

The universal characteristic of the electromagnetic field – the vector potential is related to the main parameters of the electromagnetic field by the ratio:

$$rot\,\dot{H}_m = rot\frac{\dot{B}_m}{\mu_0} = \frac{1}{\mu_0} rot\,\dot{B}_m = e_z \frac{1}{\mu_0}\left(-\frac{\partial^2 \dot{A}_m}{\partial x^2}\right) \tag{15}$$

The solution of system (14), taking into account (15) with respect to $\dot{H}_2$, is sought in the form:

$$\left.\begin{array}{ll} \dot{H}_2 = \dot{C}_4 e^{\beta x}; & -\infty < x < 0; \\ \dot{H}_2 = \dot{C}_1 e^{\lambda x} + \dot{C}_2 e^{-\lambda x} + \dot{C}_3 e^{-j\alpha x}; & 0 < x < b; \\ \dot{H}_2 = \dot{C}_5 e^{-\beta x}; & b < x < \infty \end{array}\right\}, \tag{16}$$

$$\lambda^2 = \beta^2 + \sigma^2;\ \sigma^2 = j\omega\mu_0\gamma_2 = j\alpha^2\varepsilon;\ \varepsilon = \mu_0\omega\gamma\tau^2/\pi^2, \tag{17}$$

where ε – a criterion for the quality or goodness of the device.

The method for determining the integration constants is described in sufficient detail in [5, 7], therefore, cumbersome expressions for determining the integration constants are not presented in this article.

Based on the law of the total current for the components related to the heated flat object, it is possible to write:

$$J_2 = -\frac{\partial \dot{A}_2}{\partial x} + \frac{\dot{H}_{2x}}{\delta_3'} \tag{18}$$

Taking into account (18) and (16), we obtain the formula for calculating the eddy current density in the heated plate:

$$\dot{J}_2 = \frac{\beta^2 - \lambda^2}{\lambda}\left(C_1 e^{\lambda x} - C_2 e^{-\lambda x}\right) + j\frac{\alpha^2 + \beta^2}{\alpha} C_3 e^{-j\alpha x} + \ldots$$
$$\ldots + \beta \dot{C}_4 - \frac{\beta^2}{\lambda}\left(\dot{C}_1 - \dot{C}_2\right) - j\frac{\beta^2}{\alpha}\dot{C}_3. \tag{19}$$

It contains both a harmonic component (second term), characteristic of induction heating systems that provide heating of flat products of infinite dimensions along the X axis, and additional components due to the limited size of the conducting plate.

Knowing J_2 according to (19), it is possible to determine the active specific power released per unit volume of the heated plate:

$$P_{y\partial} = J_2^2 / \gamma_2 \tag{20}$$

The force acting on the plate can be calculated, taking into account $\dot{H}_1$ according to (9), and the formula known from the theory for linear induction motors [7]:

$$F = \frac{1}{2}\mu_0 a \delta_3^{'} \int_0^b \mathrm{Re}\left[H_1 J_2\right] dx. \tag{21}$$

For flat bodies with infinite dimensions along the X coordinate, it is necessary to know the basic force per section of the heated tape with a length equal to double pole division:

$$F_{\delta a3} = \frac{\mu_0}{\pi}\tau^2 a \delta_3^{'} J_{1m}^2 \frac{\varepsilon}{1+\varepsilon^2}, \tag{22}$$

which, taking into account the additional coefficient D, which represents the ratio of the real length of the heated flat object b to 2τ, makes it possible to find the total value of the force acting on the object:

$$F = F_{\delta a3}\,\mathrm{Re}\left[(1 - j\varepsilon)\left(D - \frac{\sqrt{j\varepsilon}}{\pi(1+j\varepsilon)} \cdot \frac{ch\left(2\pi D\sqrt{j\varepsilon}\right) - \cos(2\pi D)}{sh\left(2\pi D\sqrt{j\varepsilon}\right)}\right)\right]. \tag{23}$$

Let's denote the factor in square brackets - χ. In the LIM theory, a similar coefficient takes into account the phenomenon of the transverse edge effect in the medium being moved along the Y-axis. In the general case, this coefficient is equal to:

$$\chi = \mathrm{Re}\left[(1 - j\varepsilon)\left(D - \frac{\sqrt{j\varepsilon}}{\pi(1+j\varepsilon)} \cdot \frac{ch\left(2\pi D\sqrt{j\varepsilon}\right) - \cos(2\pi D)}{sh\left(2\pi D\sqrt{j\varepsilon}\right)}\right)\right]. \tag{24}$$

As follows from the analysis of the curves shown in Figure 3, the coefficient χ depends on the quality factor ε and the ratio a/τ. Thus, with an increase of ε, the value of χ decreases, which is explained by the displacement of eddy currents from the center to the edges of the heated plate. For real induction devices, the width of the heated plates $a \geq \tau$. Therefore, an increase of χ to a certain maximum value in the range of changes for a/τ ratio from 0 to 1.0 can be neglected. In [7, 9, 10], the question of taking into account the transverse edge effect was studied quite fully. The following regularities are revealed that are of interest for the practice of induction heating:

• with an increase in slip, the effect of displacing the parameters of the electromagnetic field is more pronounced. With an increase in the speed of movement of the heated medium ($s \to 0$), the process of displacement of eddy currents from the center to the peripheral zones sharply weakens;

• weakening of the transverse edge effect is also observed in the case when the width of the plate is greater than the width of the poles of the magnetic circuit along the Y-axis. In this case, it is advisable to increase the size a only up to 40% of the value of the pole division (τ). A further increase in the size of the "protrusion" a greater than the indicated value has no practical effect on the redistribution of the parameters of the electromagnetic field and induced eddy currents.

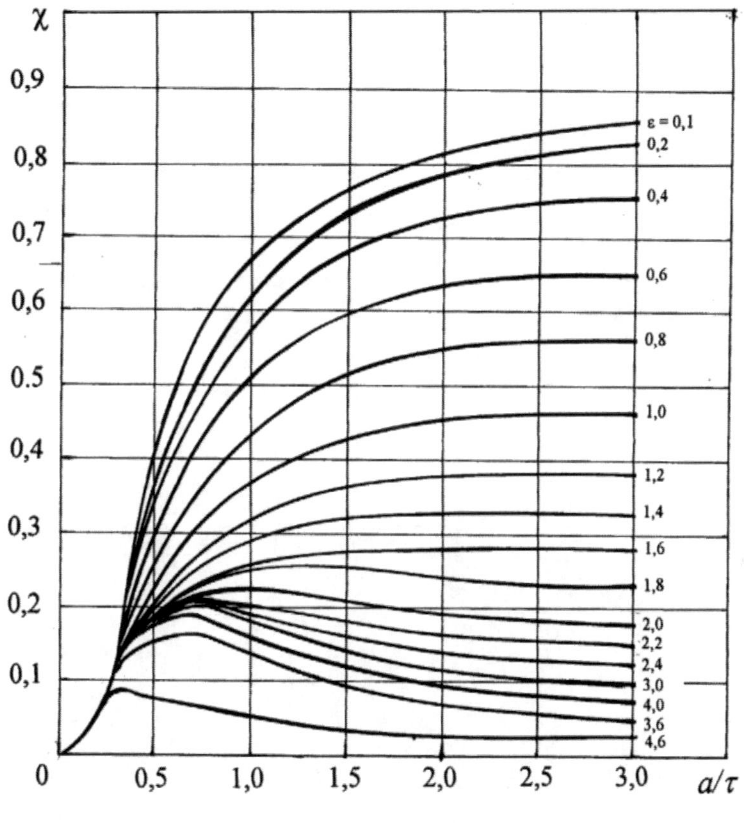

Figure 3. Dependence of the coefficient χ on a/τ and the quality criterion of the heating device.

When calculating induction devices that create a traveling electromagnetic field, it is of practical interest to solve the problem of calculating the active and reactive power released in a heated flat object. From the theory of electrical machines it is known that the total power transmitted from the primary circuit to the heated object is equal to:

$$\dot{S}_{эм} = -\dot{E}_{12}\tilde{I}_1, \tag{25}$$

where $\dot{E}_{12} = -jX_{12}I_2$ is an effective value of the EMF induced by the secondary current in the primary circuit; X_{12} is an inductive reactance of mutual induction taking into account E_{12}.

For the calculated mathematical model, the total electromagnetic power is equal to:

IOP Conf. Series: Materials Science and Engineering 1211 (2022) 012015 doi:10.1088/1757-899X/1211/1/012015

$$\dot{S}_{em} = -\int_{-\delta_3'/2}^{\delta_3'/2} dz \cdot \int_{-a/2}^{a/2} dy \cdot \int_{-p\tau}^{p\tau} \frac{\dot{E}_{12}}{\sqrt{2}} \cdot \frac{\tilde{J}_1}{\sqrt{2}} dx. \quad (26)$$

It should also be borne in mind that the induction of the secondary traveling magnetic field moves relative to the stationary windings of the inductor at a speed $\vartheta = 2\tau f_1$ and induces an EMF in them, which is equal to:

$$\dot{E}_{12} = \mu_0 \dot{H}_2 2\tau f_1 . \quad (27)$$

Using relations (2), (8) and (27), according to (26), we obtain the formula for calculating the electromagnetic power:

$$\dot{S}_{em} = S_{em\,0}\, \chi, \quad (28)$$

$$S_{em\,0} = \frac{\mu_0 p \tau \delta_3' a J_{1m}^2 2\tau f_1}{\alpha}, \quad (29)$$

where $S_{em\,0}$ is the total power in the absence of the transverse edge effect and demagnetizing action of the secondary eddy current field.

Using the relation $\alpha = \pi/\tau$, (3) и (17) for $S_{em\,0}$, according to (29) :

$$S_{em0} = 2p\tau^2 s f_1 \gamma_2 a \delta_3' \left(\frac{I_\phi m w_\phi k_{об} \sqrt{2}}{p\tau\delta_3'} \right)^2 2\tau f_1 . \quad (30)$$

Considering the above, the active (P_2) and reactive (P_{2q}) powers transmitted from the primary electrical circuit to the secondary are respectively equal:

$$\left. \begin{array}{l} P_2 = S_{em\,0}\, \chi; \\ P_{2q} = S_{em\,0}\, \chi_p, \end{array} \right\}, \quad (31)$$

where χ_p is a coefficient that allows you to separate the share of reactive power from the total electromagnetic power transmitted to the secondary circuit:

$$\chi_p = \mathrm{Im}\left[(1 - j\varepsilon) \left(D - \frac{\sqrt{j\varepsilon}}{\pi(1 + j\varepsilon)} \cdot \frac{ch\left(2\pi D\sqrt{j\varepsilon}\right) - \cos(2\pi D)}{sh\left(2\pi D\sqrt{j\varepsilon}\right)} \right) \right]. \quad (32)$$

Figure 4 shows the relation $\chi_p = f\,(a/\tau, \varepsilon)$.

It follows from the graph that the absolute value of χ_p as a function of a/τ increases monotonically for induction devices with a quality factor $\varepsilon = 0.1 \div 1.0$. For values $\varepsilon > 1.0$, the coefficient χ_p decreases monotonically.

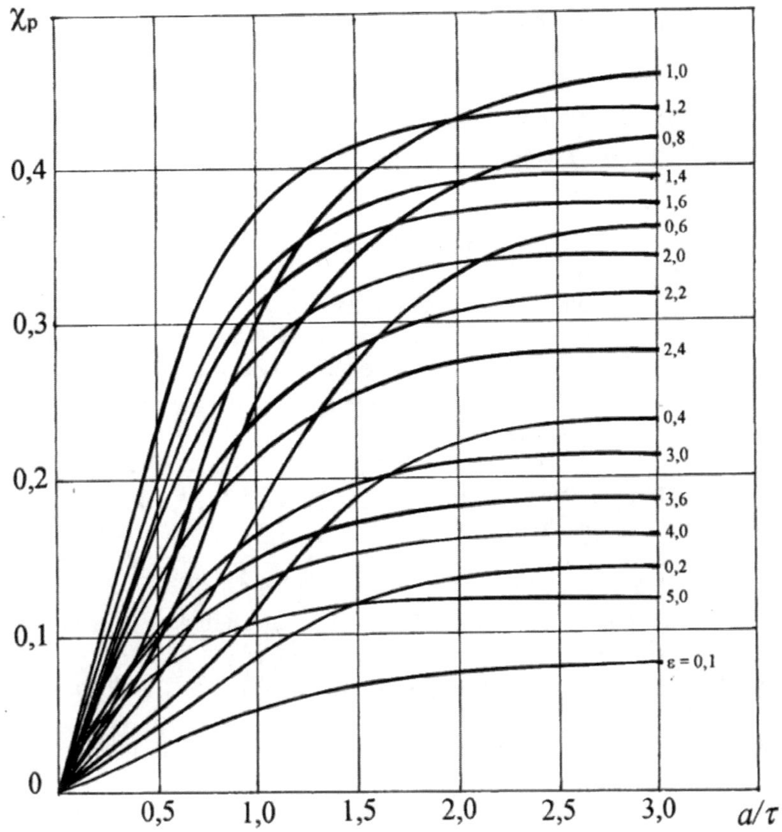

Figure 4. Dependence of the coefficient χ_p on a/τ and the quality criterion ε of the heating device.

For an idealized induction heating device, in which the active and reactive resistances of the inductor winding are equal to each other, and the transformation ratio of the EMF is equal to one, you can also write formulas for determining technical and economic indicators, which are a function of the electromagnetic quality factor:

device power factor

$$\cos\phi = \frac{\varepsilon}{\sqrt{1+\varepsilon^2}}; \tag{33}$$

device efficiency

$$\eta = 1 - S. \tag{34}$$

4. Conclusions

1. The developed mathematical model of the "inductor - heated flat metal object" system and the obtained graphical dependencies allow us to make a preliminary assessment of the efficiency of the technical indicators of induction heating devices for heating flat metal products in a traveling electromagnetic field.

2. From the analysis of the obtained mathematical expressions and the graphical dependencies given in the article, the following remark can be made. One of the main criteria for evaluating a designed or operated induction device for low-temperature induction heating of flat metal products in

a traveling electromagnetic field is the electromagnetic quality factor, which characterizes the quality of the secondary electrical and magnetic circuits of the heating device, i.e. characterizes its ability to convert energy from one type to another. So, for example, the greater the value of the product of quality factor (ε) and slip (s), the more energy consumed is converted into mechanical energy, thereby reducing the heating efficiency.

5. References

[1] Kobe E, Kuvaldin A B 1982 Induction heating *THI Section Electrical Engineering and KPD* (University section) p 110

[2] Kuvaldin A B 1998 Induction heating of ferromagnetic steel (Moscow: Energoatomizdat) p 200

[3] Kachanov A N, Kachanov N A 1988 Application of low-temperature induction heating systems in peasant and farm farms of the Republic of Kazakhstan // Recommendations. TSNTR under the Akim of Pavlodar region and the Ministry of Science-Academy of Sciences of the Republic of Kazakhstan (KazgosINTI) p 40 p

[4] Kachanov A N , Kachanov N A, Korenkov D A 2016 Classification and scope of low-temperature heating systems with open magnetic circuits *Bulletin of the Moscow Power Engineering Institute* **2** 36-40

[5] Voldek A.I. 1970 Induction magnetohydrodynamic machines with a liquid-metal working fluid. (L.: Energiya) p 270

[6] Krumin Yu K 1983 Fundamentals of theory and calculation of devices with a traveling magnetic field (Riga: Zinatne) p 184

[7] Veselovsky O N, Konyaev A Yu, Sarapulov F N 1991 Linear asynchronous motors (M.: Energoatomizdat) p 256

[8] Nuns J, Poiroux R 1981 Le chauffage de bandes d'aluminium par inducteur a flux transverse *E.D.F. Bulleten de La directions des etudes et Recherches – Serie B reseaux Electrigues, Materiels Electr*igues **3** 57 – 94.

[9] Yamura S 1983 Theory of linear asynchronous motors. // Translated from English (L.: Energoatomizdat) p 160

[10] Kachanov A N , Kachanov N A, Golda AV 2012 Calculation of the distribution of the main parameters of the electromagnetic field in a metal load with unilateral heating by induction heating *Fundamental and applied problems of engineering and technology* **6** 35-40

Energy-efficient omd processes with complex local loading for the production of machine parts with a given level of properties

V A Golenkov, D O Dorohov and S Y Radchenko

Orel State University, 95, str. Komsomolskaya, Orel, 302026, Russian Federation

Abstract: The article discusses some new energy-efficient metal forming processes (PMD) with complex local loading of the deformation zone for the production of machine parts with a given level of properties. The place of such processes in the aggregate of all OMD technologies is shown, and their classification is given. The initial data of numerical mathematical modeling of the specified class of processes for a particular case are given, the features of the stress-strain state of the metal in the deformation zone are revealed. In the study, from a theoretical point of view, one of the important practical aspects of these technologies is considered: the effect of the value of axial compression on the shape and size of the deformation zone arising from the action of local loading of the workpiece.

1. Introduction

Advancements in mechanical engineering are raising the bar ever higher. Today, any new casting, cutting, or forming technology is expected to surpass its predecessor as well as to compete against additive technologies. The authors hereof develop and implement technologies based on complex local loading of the deformation zone (CLL deformation) as part of the broader metal forming research. CLL deformation represents a set of metal forming processes, which create a plastic deformation zone by applying multiple loads, one of which affects the bulk of the workpiece and is referred to as global loading; the rest is local and is intended to generate a non-stationary deformation zone characterized by a complex stress state. [1-10]

2. Statement of Problem

Before delving deeper into the problem, let us define the complex and local loads. Complex loading (CL) is the process of applying two or more independent forces and/or moments to a workpiece. Torsional tensile tests are a classic example. Apparently, CLL deformation imposes a complex load upon the workpiece material [4-8]. This infers the following definition: CL is a process that subjects a workpiece to two or more independent loads, which entails a complex non-homogeneous stress state in the deformation zone.

If this or that metal forming technology implies deliberately localizing the deformed material in a non-stationary plastic flow zone that is far smaller than the bulk of material deformed by a single operation, this is a local loading (LL) process. Spacing operations that deform a sheet, e.g. bending or flanging, are non-LL, as they cause plastic deformation limited to a particular location, i.e. unlike in drawing, the instant deformation zone does not move along the workpiece. Pressing and rotational forging processes are not local either; although the instant deformation zone is limited in scale, the process deforms the bulk of the workpiece to transfer pressure. Thus, local deformation only comprises the processes that limit active deformation to a specific spot that moves along and/or within the workpiece. Hence the definition: local loading is a process that is intended to create a local active non-stationary deformation zone.

Let us present the general CLL deformation diagram to better describe the concept. The workpiece is globally loaded by a system of forces and moments shown in Figure 1 as a distribution of stresses f. The workpiece is also locally deformed by the tools 1, 2, … n that follow the trajectories Ω_1, Ω_2, etc. The tools localize the deformation zone to intensify the plastic flow in a small part of the workpiece.

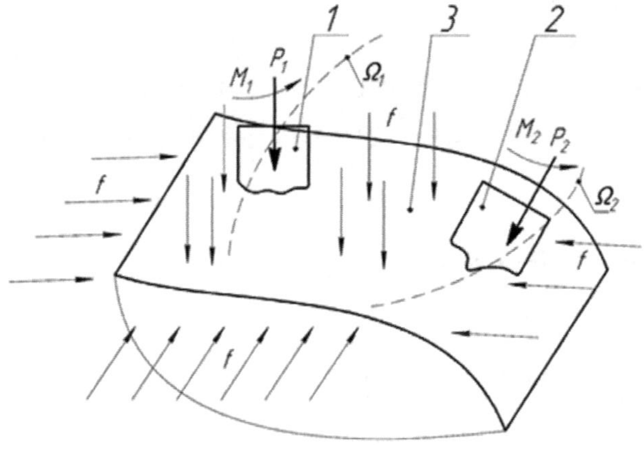

Figure 1. CLL deformation: general diagram
1, 2 are local deformation tools; 3 is the workpiece; P_i, M_i are tool-specific forces and moments; f is the stress caused by global loading forces; Ω_i are tool-specific trajectories.

As the tools and the workpiece are displaced with respect to each other, the deformation zone moves. As a result, one part of the workpiece is subjected to single or multiple loading for more efficient reshaping and/or hardening [1-10].

From the standpoint of machining goals, CLL deformation processes can be systematized as follows [1-5].

1. Reshaping CLL deformation that seeks to alter the shape; hardening may occur as a not necessarily desirable side effect. Reshaping can be mainly attained by:
- global loading, in which case local loading is auxiliary;
- local loading, in which case global loading is auxiliary and serves to improve the stress state diagram. As such, it can be imposed indirectly by form locking;
- a combination of global and local loading.

2. Hardening CLL deformation, in which case global loading is auxiliary and serves to improve the stress state diagram. As such, it can be imposed by:
- applying a static load that does not cause any stress equal to, or exceeding the yield point;
- indirectly by form locking.

3. Combined CLL deformation, which seeks to reshape the material while also improving its structure and the product strength. As such, it combines active global and local loading.

Diagrams 1.2 and 2.1 are preferable from the standpoint of energy conservation and ease of use. These are processes where the global loading component serves to adjust the stress state diagram and does not cause any stress that would equal or exceed the yield point of the workpiece material. To analyze these CLL deformation processes, dissect the general CLL deformation diagram with a plane that passes through the forming tool and the smoothing tool, see Figure 2.

Distributed load can be applied in various ways, see Figure 3. When implemented, Option *a* implies that the workpiece move along the supporting surface; at $p_1 = p_2$, the workpiece travel speed is constant, no stress emerges in the deformation zone, which means this option is useless. At $p_1 < p_2$ or $p_1 > p_2$, the workpiece movement is associated with additional forces emerging in the deformation zone under the tools 1 and 2; however, the magnitude of such forces is the difference of p_1 and p_2, making machining less efficient. Thus, Option *a* is hardly usable.

With Option *c* as shown in Figure 3, the distributed loads p_1 and p_2 cause tensile stresses in the workpiece body; with Option *b,* the stresses are compressive. Thus, there are four practically applicable global loading options for CLL deformation in processes where global loading is intended to alter the stress-state diagram, see Table 1.

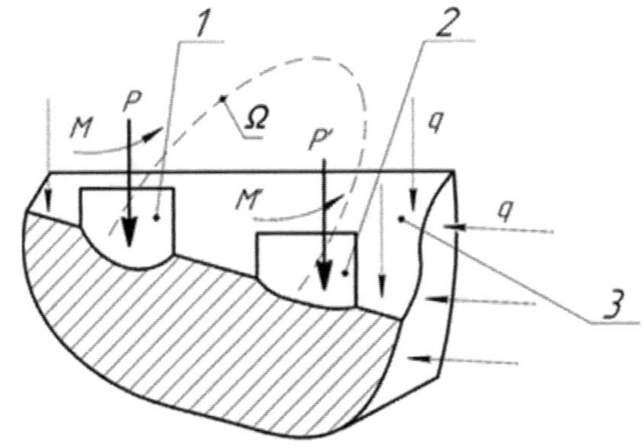

Figure 2. CLL deformation: general diagram section.

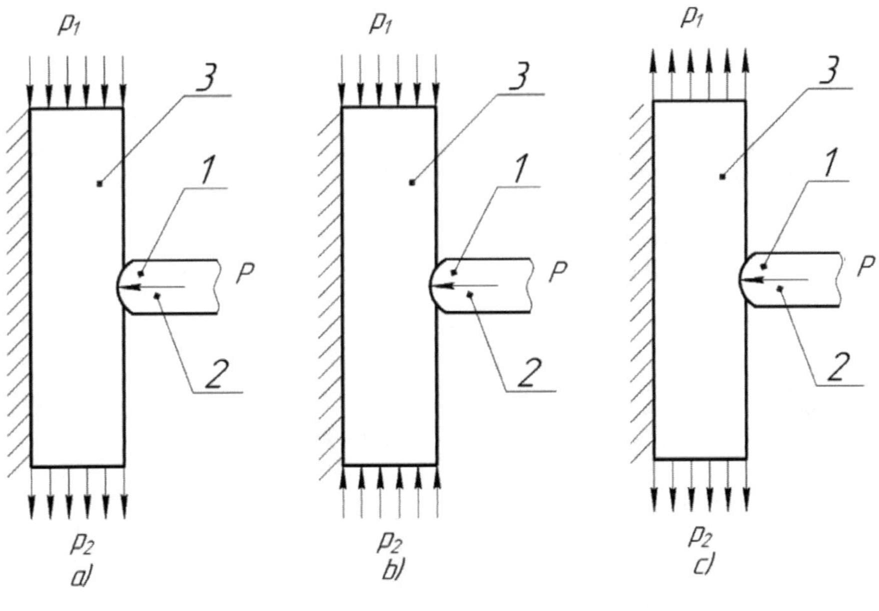

Figure 3. Applications of distributed load: 1 and 2 are local deformation tools (shown as one for simplicity); 3 is the workpiece.

Apparently, the values p_1 and p_2 can significantly affect the plastic flow of metal when machining. It is therefore important to analyze how compressive and tensile forces affect the stress-strain state in such CLL deformation processes where such forces do not cause any stress that would exceed the yield point.

Table 1. Practical CLL deformation technologies where the global load does not cause any stress in excess of the yield point.

Global load application options	p_1 & p_2 ratio	
	$p_1 = p_2$	$p_1 < p_2$
b	hardening CLL deformation	-
c	Pull-backpull rolling: particular case	Pull-backpull rolling: general case

Let us assess the effects of axial compression in case of hardening CLL deformation, i.e. Option *b* at $p_1 = p_2$. Tests are based on mathematical modeling in Stamp package [2].

The problem-solving algorithm is based on a mathematical model that uses a well-known variation principle:

$$\iiint\limits_{V(t)} \delta\nabla \mathrm{v}\cdot\cdot({}^{r}\sigma - \mathrm{d}\cdot\sigma - \sigma\cdot\omega + \mathrm{w}\cdot\sigma - \sigma\cdot\mathrm{w} + (\nabla\cdot\mathrm{v})\sigma)dV = \\ = \dot{\mathrm{Q}}\cdot\delta\dot{\mathrm{q}} + \iint\limits_{S(t)} \delta\mathrm{v}_t\cdot(\dot{\mathrm{p}} + \mathrm{p}(\nabla\cdot\mathrm{v} - \mathrm{n}\cdot\mathrm{d}\cdot\mathrm{n}))dS , \quad (1)$$

where σ, d, ω are the Cauchy stress tensor, the strain-rate tensor, and the rotation-rate tensor, respectively.
p, n are the pressure and the external normal relating to the actual contact surface $S(t)$,
$\dot{p}$ is the pressure rate,
r is the co-rotational derivative in the constitutive equation for the material of the deformed solid,
$\nabla \cdot \mathrm{v}$ is the velocity field gradient,
$\dot{\mathrm{Q}}$ are generalized-force velocities,
$\delta \mathrm{v}_t$ are variations of the velocities of sliding along the surface of an absolutely rigid solid,
$\delta\dot{q}$ are variations of generalized velocities.

Integration (1) by the volume $V(t)$ and the contact surface $S(t)$ at time t.

The equation (1) is a modified velocity variation principle proposed in 1983 by L.A. Tolokonnikov, O.L. Tolokonnikov, A.A. Markin and V.F. Astapov, later advanced by P.G. Morev [2-5].

The velocity field gradient v(y) is defined as:

$$\nabla \mathrm{v} = \mathrm{э}^i \frac{\partial \mathrm{v}}{\partial y^i} = \nabla_i \dot{y}^j \mathrm{э}^i \otimes \mathrm{э}_j = \nabla_i \dot{y}_j \mathrm{э}^i \otimes \mathrm{э}^j ; \quad (2)$$

The gradient of the velocity field v(y) is defined as:

$$\nabla \mathrm{v}^T = \frac{\partial \mathrm{v}}{\partial y^i}\mathrm{э}^i = \nabla_i \dot{y}^j \mathrm{э}_j \otimes \mathrm{э}^i = \nabla_i \dot{y}_j \mathrm{э}^j \otimes \mathrm{э}^i ; \quad (3)$$

the strain-rate tensor is defined as:

$$\mathrm{d} = \frac{1}{2}(\nabla \mathrm{v} + \nabla \mathrm{v}^T), \quad (4)$$

the rotation-rate tensor is defined as:

$$\omega = \frac{1}{2}(\nabla \mathrm{v}^T - \nabla \mathrm{v}), \quad (5)$$

the velocity field of the material points of the solid determines their law of motion

$\dot{\mathrm{y}}(\mathrm{y},t) = \mathrm{v}(\mathrm{y},t)$.

The following ratio is used as the constitutive equation for the strain-rate tensor *d* and the stress tensor σ

$${}^{j}\sigma = \mathrm{D}\cdot\cdot\mathrm{d}, \quad (6)$$

where *j* is Yauman's derivative,
whereas tensor D is written as follows in global Cartesian coordinates:

$$D^{ijkl} = \begin{cases} E^{ijkl} - \dfrac{3GS^{ij}S^{kl}}{\bar{\sigma}^2(1+H'/(3G))} & \textit{under active loading} \\ E^{ijkl} & \textit{under reversible loading} \end{cases} \tag{7}$$

where E^{ijkl} is the isotropic tensor of elasticity,

S^{ij} is the stress-tensor deviator,

$\bar{\sigma}$ is the stress intensity,

H' is the hardening modulus,

G is the shear modulus.

Here is the condition of elastic-to-plastic state transition:

$$\bar{\sigma} = H(q), \tag{8}$$

where $H(q)$ is the hardening curve.

Mathematical modeling relies on the isotropic theory of plasticity with low elastic deformation, as hardening uses cyclical or nearly cyclical workpiece deformation. Odqvist parameter increments are rather small for each cycle. Therefore, the reversibility surface will not differ much from a circumference, hence no contradiction to the actual deformation. The evolutionary equation of stress state is then written as follows:

$$\dot{\sigma} = {}^{j}\sigma - \sigma \cdot \Omega + \Omega \cdot \sigma, \tag{9}$$

where Ω is the rotation speed of the material point neighborhood.

To solve the equation (2) numerically, discretize its left side and use the constitutive equation written as:

$${}^{r}\sigma = \mathrm{D} \cdot\cdot \nabla \mathrm{v}. \tag{10}$$

The mathematical model uses Coulomb's isotropic law with a constant coefficient of friction μ that links the tangent pressure component p_t and the normal pressure component p_n:

$$|\mathrm{p}_t| \le \mu \cdot |\mathrm{p}_n|. \tag{11}$$

The mathematical model describes the quasi-static interaction of rigid solids with a finite elastoplastic solid. Finite-element analysis is used to solve the system of reduced equations (1), (6)-(11).

Figure 4 presents the calculation diagram for mathematical modeling. Calculation uses the following boundary conditions:

- nodes in contact with the indenter may not move inside the tool;
- nodes in contact with the hatching may not move at all.

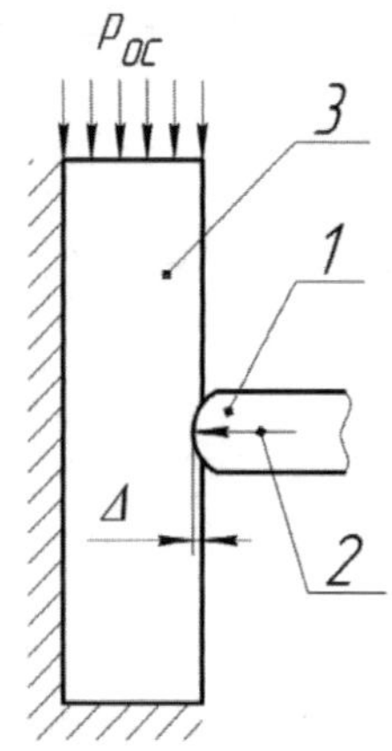

Figure 4. Calculation diagram of a single roller injection when CLL deformation is used for hardening: 1 is a deforming roller (indenter); 2 is the trajectory of single roller injection to a depth, 3 is the workpiece, 4 is the symmetry axis.

The injectable body is a toroidal roller, 0.5 mm in radius. To simulate the power action of the roller, use relative injection depth defined against the workpiece thickness. Find the axial compression as a fraction of the yield point for the material under consideration $\sigma_T = \sigma_0$.

3. Results

Figure 5 shows the results of modeling the injection of a toroidal tool into the workpiece body to a relative depth of 0.02 at different axial compression magnitudes. These are Odqvist parameter isolines at zero. Thus, what is shown in Figure 5 effectively defines the hardened zone of the workpiece. Figure 5 clearly shows that in case of axial compression, the hardened zone for a toroidal roller splits into two petal-shaped zones. Note that similar phenomena occur with other tools, too, see Figure 6. As shown in Figure 6, the applied axial force must exceed $0.9\sigma_0$ for the hardened zone to grow.

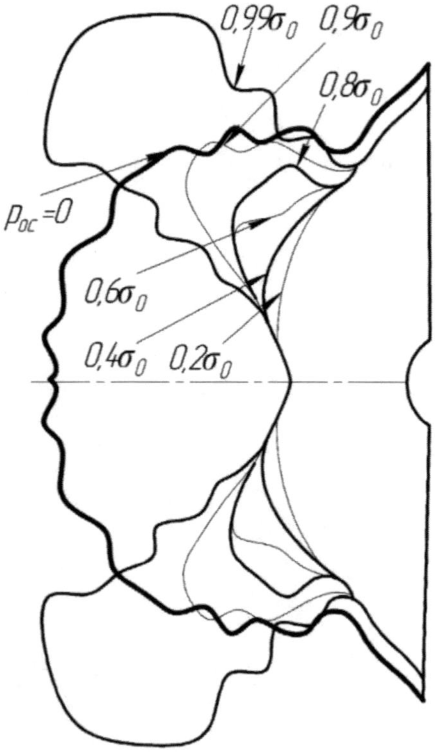

Figure 5. Injecting a toroidal roller: fluctuation in the shape of the hardened zone as a function of axial compression.

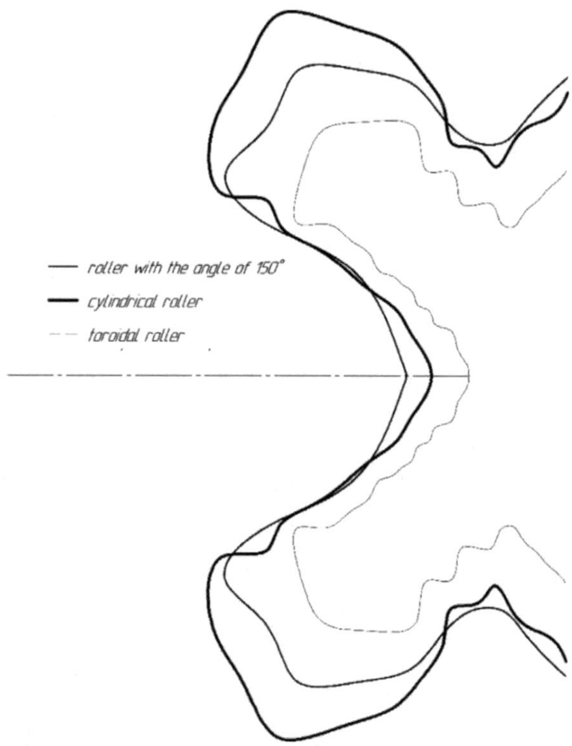

Figure 6. Comparing the geometry of hardened zones for three tool types at $p_{oc} = 0.99\sigma_0$ and $\Delta_r = 0.020$

Table 2 presents data on increasing the relative hardening depth and peak Odqvist parameter value at maximum vs zero axial compression. Apparently, the hardening depth and magnitude peak when applying extra axial compression at low injection depth. It should be noted that maximizing the Odqvist parameter is not necessarily desirable, as it can destroy the product [1, 6-10].

Table 2. Relative growth of Odqvist maximum value and relative depth of hardening at different depth of implementation and geometryof indentors for $p_{wasp\ valuescompared}$ 0 and p wasps $0.9to_0$.

relative injection depth	0.002	0.004	0.008	0.012	0.016	0.020
increase in the peak Odqvist parameter value, %	22.02	20.24	19.12	16.44	16.09	15.44
increase in the relative hardening depth, %	144.09	72.71	41.16	17.59	6.50	1.93

4. Experimentation

Based on the diagram shown in Figure 1, the research team developed a process, see Figure 2, that effectively performs complex local loading of the deformation zone for the purpose of hardening. The global load application option for this method is *b* at $p_1 = p_2$. The workpiece in use was the bush 1, see Figure 2. The local deformation tool was the special lugged roller 5. Like the roller, the roller matrix 4 was used for local deformation. The workpiece was driven by stops through a rigid geometric connection. Machining was performed in the three-roller matrix 4. Hardening started at the outer surface; the inner surface rested on the frame 3, which controlled and constrained the plastic flow of metal.

It's no coincidence we selected a bush for experimental CLL deformation, as most parts that require improved mechanical properties are axisymmetric (shafts, axles, sleeve bearings, piping products). Should be noted that there's no fundamental difference between machining bushes as shown in Figure 7, pipes, axes, and shafts. Using rollers for local loading was also an obvious solution that reduced friction losses. The lug had to geometrically match the process diagram and feature a symmetry axis, as the tool moved reciprocally.

The significant process parameters were determined following the method description (Figure 7) and the general diagram (Figure 1).

1. Geometry and shape of the forming tool for the roller 5.

2. Number of passes, i.e. the number of closed trajectories Ω_1 the forming tool 1 follows when traversing the workpiece surface. For the technology shown in Figure 2, it corresponds to the number of roller passes along the workpiece.

3. Loading force (the tool injection depth Δ) is the force P applied to the roller.

4. Forming tool feed pitch (s in Figure 7) is a parameter of the trajectory Ω_1. When machining an axisymmetric product, the trajectory Ω_1 is a helix. The workpiece 1 in Figure 2 rotated around its axis while the roller 5 moves in parallel to this axis at the speed υ, which meant the feed pitch was the helix pitch.

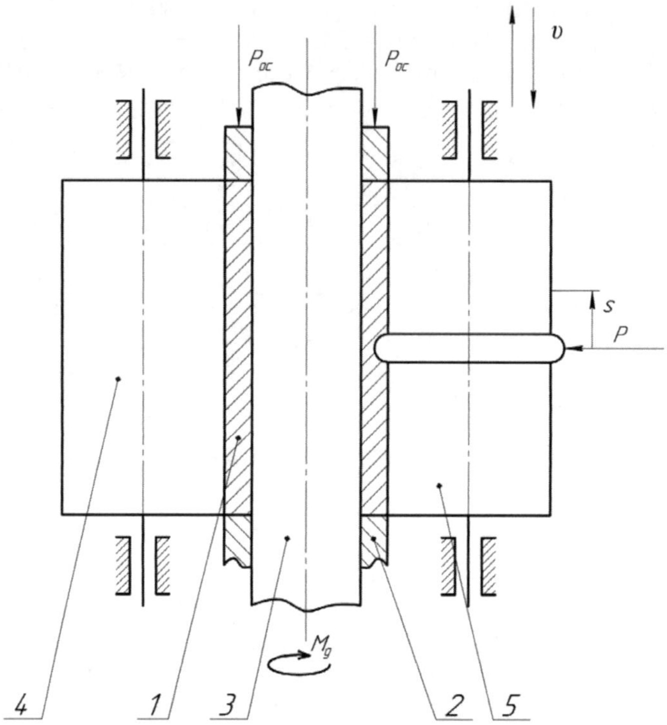

Figure 7. Hardening CLL deformation diagram:
1 is the workpiece; 2 are stops; 3 is the frame; 4 is the roller matrix (a smoothing tool); 5 is the lugged roller (a forming tool); M_g is the moment applied to the workpiece; υ is the roller travel speed; P is the force applied to the roller to control the injection depth; P_{oc} is the force applied to the stops, s is the feed pitch.

The smoothing tool (Figure 7) restores the surface to its original geometry after the forming tool had passed. Global loading had the force P_{oc}, which for these experiments was form-locking, i.e. $p_1 = p_2 = 0$. Experiments for cases when $p_1 = p_2 > 0$ are work in progress. In case of force closure, the force P_{oc} should not cause any stress in the workpiece body equal to, or exceeding the yield point provided zero sundry load, i.e.

$$P_{oc} < \sigma_T \cdot \frac{\pi \cdot \left(D^2 - d^2\right)}{4}, \tag{12}$$

where D is d are the workpiece inner/outer diameter.

Experiments used specially designed tooling that could vary the force (the roller injection depth), the tool geometry, the number of passes, and the feed pitch. Processing the experimental data produced the distribution curves of microhardness (H_μ) as a function of distance to the surface (h_μ, μm) and number of passes (n) at a certain tool force (P, H), see Figure 8 as an example [8-10].

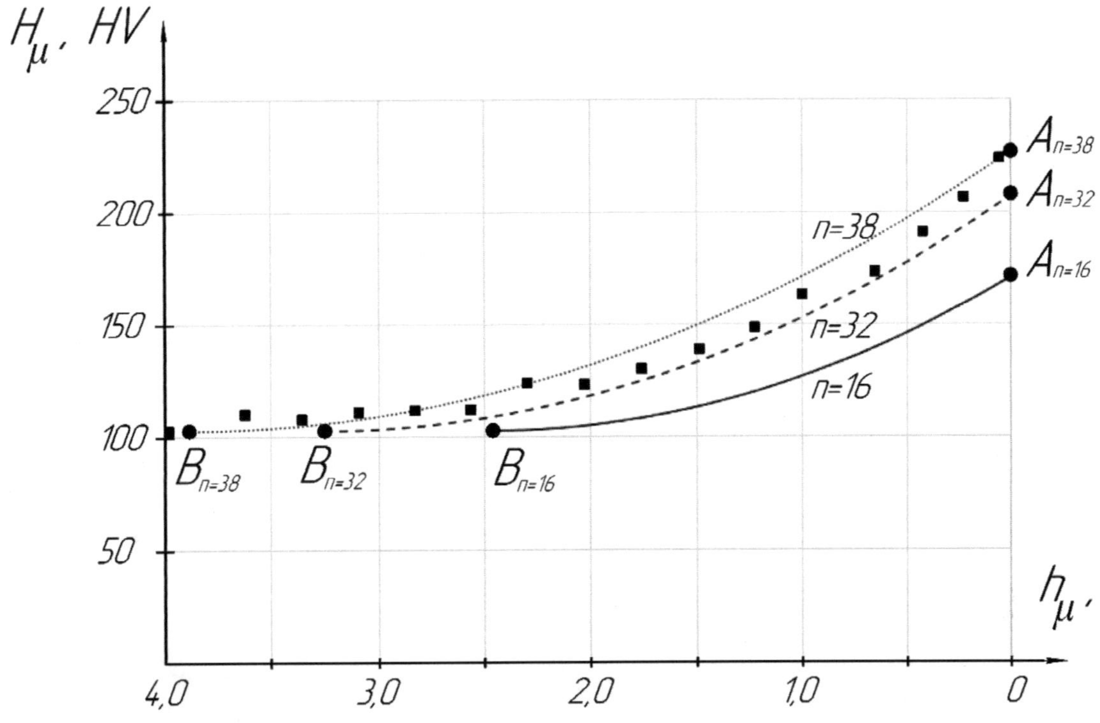

Figure 8. Distribution of microhardness values in the coordinates $H_\mu - h_\mu$ for the hardening CLL deformation at different pass numbers and P = 310N; points A are surficial, points B are located at the interface of hardened/non-hardened areas (experimental data approximated by the least squares). Squared dots show the experimental distribution of microhardness parameters for n = 38 and P = 310N.

5. Discussion

The hardened zone features pronounced "petals" when exposed to axial compression at $p_{oc} > 0.9$. A closer look at the stress distribution in the deformation zone (Figure 9) sheds some light. As shown in the figure, zones featuring σ_{xr} and σ_{zr} (values of stresses σ_x and σ_z normalized to σ_0) are similar in size and geometry. However, the entire workpiece is stressed in the longitudinal section for $p_{oc} = 0.99\sigma_0$, see Figure 9. This is due to the fact that in this plane, workpiece parts are closer to a transition to plastic state prior to tool injection than in any other plane as they are exposed to axial compression. Thanks to these "petals", the hardened zone actually splits into two components.

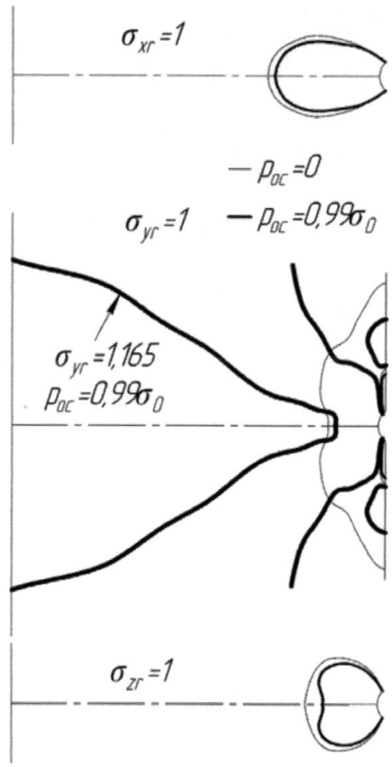

Figure 9. Relative compressive stress isolines showing the plastic/elastic deformation interface for the toroidal roller at axial compression $p_{oc} = 0.99\sigma_0$ or no axial compression, relative injection depth = 0.02.

This is a significant advantage of the methods that apply extra load. Given that hardening CLL deformation is associated with longitudinal feed at a certain pitch, the petals to the left and to the right of the roller path will inevitably intersect. Apparently, each individual petal will, all other things being equal, pass the same portion of the workpiece twice, which means applying an axial load $p_{oc} > 0.99\sigma_0$ makes machining twice as efficient as applying zero axial compression.

6. Conclusions

The paper considers metal forming by local loading of the deformation zone. The method is described and classified as a part of the metal forming technology. Theoretical analysis covers one of the most important aspects of this technology, i.e. how axial compression affects the geometry of the deformation zone created by locally loading the workpiece. The paper shows that applying axial load in machining is appropriate as long as the load is 90% to 99% of the yield point when CLL deformation is used for hardening. It is also shown that all other things being equal, axial compression can double the efficiency of machining.

REFERENCES

[1] V.A. Golenkov, S.Yu. Radchenko, D.O. Dorokhov, and G.P. Korotkii, "Scientific Foundations of Strengthening via the Combined Local Deformation", Moscow: Mashinostroenie; Oryol: State University, UNPK, 2013

[2] Yakovlev, S.S., Tregubov, V.I., Osipova, E.V. Limiting deformation in rotary drawing of anisotropic pipe blanks with wall thinning *Russian Engineering Research* 36(6), 2016, 472-475

[3] Yakovlev, S.S., Polikarpov, E.Yu. Extension of thick-walled cylindrical anisotropic blanks, with thinning of the wall. *Russian Engineering Research* 29(7), 2009, 698-704
[4] Morrev, P. G. (2011). A Variational Statement of Quasistatic “Rigid-Deformable” Contact Problems at Large Strain Involving Generalized Forces and Friction. Acta Mech, 222, 115-130. http://dx.doi.org/10.1007/s00707-011-0516-9
[5] Morrev, P. G., Fedorov, T.V. A nurbs approximation of experimental stress-strain curves. *Journal of Chemical Technology and Metallurgy*. 2016, 51(3). 341-349.
[6] Fedorov, T.V. Approximation of experimental hardening curves by inhomogeneous fractional rational B-splines [Text] / T.V. Fedorov // *Fundamental and applied problems of engineering and technology*. 2014. № 1 (303). 64-68.
[7] Morrev, P.G., Gordon, V.A. Simulation of surface hardening in the deep rolling process by means of an axial symmetric nodal averaged finite element. *Journal of Physics: Conference Series*, 2018, 973(1),012013.
[8] Morrev, P.G., Gordon, V.A. An axisymmetric nodal averaged finite element. *Latin American Journal of Solids and Structures*. 2018. 15(2),e14
[9] Pilipenko, O.V., Radchenko, S.J., Golenkov, V.A., Dorohov, D.O. Numerical Mathematical Simulation of Penetration of Indenters of Different Shapes During Strengthening Ma-chine Parts via Local Loading of Deformation Zone // International Symposium on Engineering and Earth Sciences (ISEES 2018). Advances in *Engineering Research*, volume 177, 568-573.
[10] Pilipenko, O.V., Radchenko, S.J., Golenkov, V.A., Dorohov, D.O. Gradient Strengthening Control Procedure Based on Numerical Simulation of Combined Local Loading Of Deformation Zone // International Symposium on Engineering and Earth Sciences (ISEES 2018). Advances in *Engineering Research*, vol. 177, 574-577.

Automated fire risk evaluation of electrical installations in the man-machine system

M A Gabova[1], O K Nikolsky[2], Yu D Shlionskaya[2]

[1] Department of Information Systems in Economics, Polzunov Altai State Technical University, 46, Lenina ave., Barnaul, 656038, Russia
[2]Faculty of Energy, Polzunov Altai State Technical University, 46, Lenina ave., Barnaul, 656038, Russia

Abstract. The article considers approaches to the formation of a system of criteria for assessing the electrical installations fire condition of the agricultural and industrial complex. Based on the analysis of the literature, the conclusion is made about the appropriateness of the use of expert assessments. To implement the decision, a group of experts was assembled, on the basis of whose knowledge a list of 42 parametersб characterizing the fire condition of the electrical installation was determined. To identify the relationships and form a method for calculating the estimated value of fire risk, experts assessed the fire condition of 70 electrical installations of the agricultural and industrial complex of the region. A knowledge base was formed from the resulting values. As a method of data analysis, it was decided to use neural networks, but the available sample is not sufficient for high-quality training of a neural network. Therefore, the correlation method and the principal component method were considered, and based on the calculations, it was decided to use a training sample consisting of 6 principal components for training a neural network. A neural network was trained on these data and the values of the average error were obtained sufficiently low, which may indicate sufficient accuracy of the generated model. The article also presents a conceptual scheme of a software package for automating calculations in accordance with the developed model.

1. Introduction

The operational safety state of electrical installations in various sectors of the economy (agriculture, electric power, communal services) has reached a critical level, which, of course, poses a threat to national economy.

The most dangerous is the fire situation, 30% of the causes of fires occur for electrical reasons.Moreover, the number of fires, losses (deaths and injuries of people) and material damage, as well as significant deterioration of the environmental situation tend to increase significantly both in the short and long term.

Governments regulation mechanisms of electrical installations fire safety in buildings have shortcomings. Insufficient enforcement of regulations and laws leads to fires [1].

According to the Ministry of the Russian Federation for Civil Defence, Emergencies and Elimination of Consequences of Natural Disasters about 1/3 of all fires occur for electrical reasons [1].

The current situation poses a threat to the national security of the country, and also indicates the need to improve fire safety systems based on scientifically sound solutions.

One of the main reasons for the existing negative technogenic situation in the fire sector is the insufficient elaboration of the problem of electrical installations ensuring fire safety.

In order to improve the current situation, it is necessary to develop practical recommendations based on the analysis of factors that are the causes of fire hazard risks of electrical installations in the agro-industrial complex.

To reduce the number of accidents leading to fires at electrical installations and electrical injuries, large amounts of data are collected and processed, which entails the need for automation of these processes.

Work in this direction was carried out in [2,3], but, these works leave open questions about the use of intelligent technologiesto manage the fire risks of electrical installations and create simulation models based on them to form recommendations for preventive measures, that reduce the fire risks of electrical installations.

Reducing the number of fires and damage from fires is possible only by carrying out preventive measures to inspect the current state of electrical installations and implementing measures to prevent fire-hazardous situations.

2. Methodology of Assessment and Management of Agricultural Objects Technogenic Safety

Many Russian and foreign scientists have been dealing with the risks of operating electrical networks with a voltage of up to 1000 V, the results of their work have been the creation of theoretical foundations of electrical and electrical installations fire safety, the formation of methodological and regulatory frameworks on its basis for the creation and operation of safe exploitation electrical installations.

To determine the risk level of various factors, many authors use mathematical modeling based on probabilistic analysis the risks of operating electrical installations and the development of practical calculation methods [4]. In order to test the developed models, it is necessary to collect data on the condition of real objects, which begins with technical diagnostics of the fire condition, consisting of three elements: control, measurement and testing.

Subsequent presentation of the data obtained should be carried out in accordance with the concept of the human-machine system "Human - Electrical Installation - Environment" [5].

The issues related to the analysis and synthesis of a weakly structured dynamic human-machine system "Human-Electrical Installation-Environment" (H-E-S), which allows identifying risk factors, are sufficiently studied [6, 7], however, the question of choosing the most optimal calculation method remains open.

Fire risk management is based on decision-making in conditions of uncertainty associated with such features as:

a) lack of a clear structure, uncertainty and physical heterogeneity of the system components;

b) difficulty in identifying events leading to the emergence of dangerous man-made situations;

c) the complexity of obtaining and interpreting data describing the state of the electrical installation.

These factors, as well as the lack of an accurate and complete description of the relationships between the values of indicators, the occurrence and development of emergency situations, complicate the decision-making procedure.

To solve the problem of increasing the diagnostics efficiency of the electrical installations technical condition, it is necessary to create a new methodology based on algorithmization of technogenic risk management throughout the entire life cycle of an electrical installation of a production facility.

The aim of the work is to reduce fire risks of electrical installations, based on mathematical models and automate calculations with the use of intelligent components for analyzing diagnostic data on the condition of electrical installations.

The objectives of the work are:

- identification of parameters that have the greatest impact on the level of fire risk of electrical installations;
- conducting experiments to select the most optimal calculation methods;
- development of a software package that automates calculations.

The quantitative indicator that most fully characterizes the technical condition and residual life of electrical installations is the fire risk Rf» [8].It is known that the basis for an objective assessment of the risks of an electrical installation can be field experiments, the conduct of which in some cases is unacceptable for humane or economic reasons [9]. The determination of quantitative and qualitative indicators of the state of an electrical installation is fraught with difficulties both in obtaining and interpreting them. That’s why to solve the problem was used the expert evaluation method

IOP Conf. Series: Materials Science and Engineering 1211 (2022) 012017 doi:10.1088/1757-899X/1211/1/012017

3. Methodology of assessment of agricultural objects fire safety based on human-machine system "H-EI-E".

To establish the parameters, characterizing the fire condition of the electrical installation, was formeda group of experts from 6 people, specialists in electrical engineering and fire safety.They determined the parameters, that influence greatest impact on the electrical installationfire risk. The results obtained in accordance with the concept of the human-machine system "Human-Electrical Installation-Environment" were divided into 3 clusters and presented in the form of a hierarchical structure (Figure 1.)

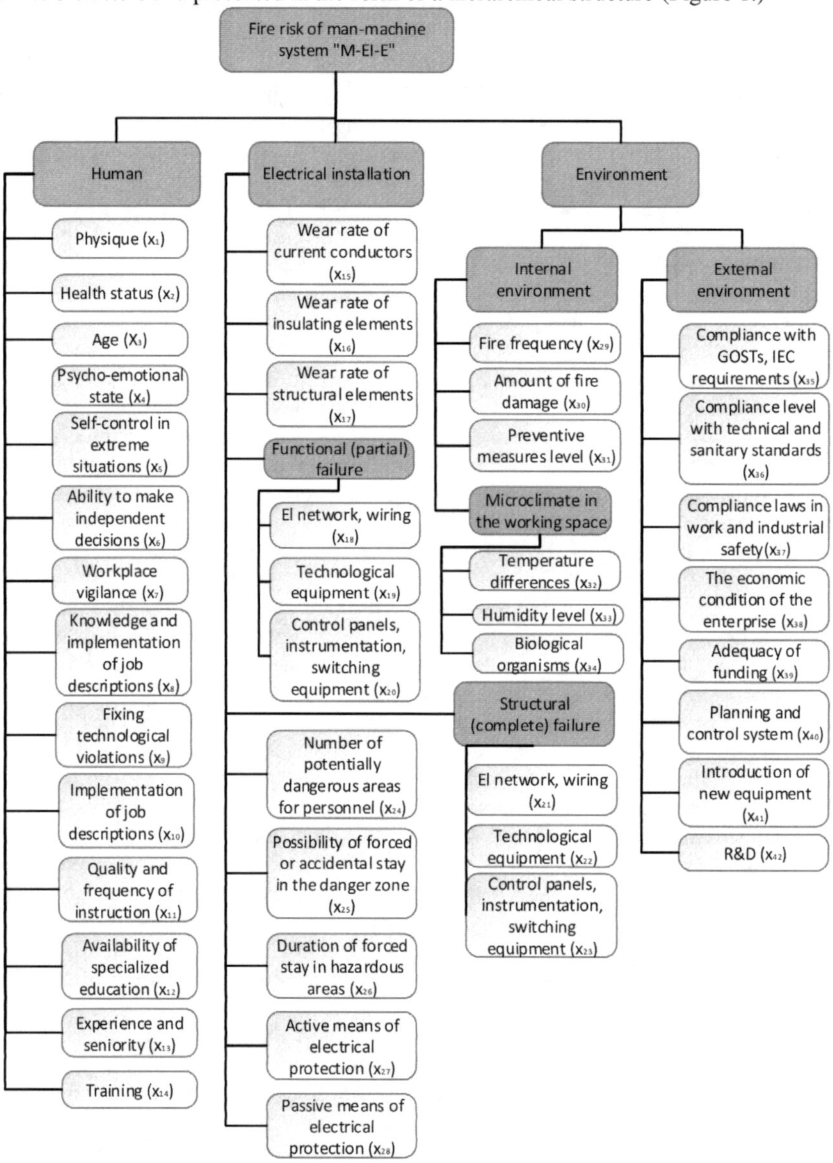

Figure 1. Hierarchical diagram of fire risk-forming factors of human-machine system "H-EI-E"

Experts identified 42 parameters, which were divided into three clusters:

"Human" is the content of data on the physical, psychoemotional state of the facility's personnel, their professional knowledge and work experience. "Electrical installation" - assessment of the technical condition of the electrical installation itself, its structural and functional elements, applied protective equipment and fire-hazardous areas on or near the object.

"Environment" is divided into two subsections:

"Internal environment" - characterizing the frequency of fires, the level of preventive measures and the microclimate in the work space, where the electrical installation is operated.

"External environment" - the level of compliance with laws, GOST standards, technical and sanitary standards, as well as a number of economic indicators characterizing the general condition of the enterprise, updating outdated equipment and conducting research and development.

Each of the parameters takes a value in the range 1-3. Which corresponds to the fire risk level defined in GOST 12.1.004-91, where the permissible level of fire danger for people should be no more than 10^{-6} exposure to fire hazards exceeding the maximum permissible values per year per person.

3 – acceptable risk ([<=1*(10^{-6})]);

2 – tolerable risk ([1*(10^{-4}-10^{-5})]);

3 – unacceptable risk ([1*(10^{-2}-10^{-3})]).

In accordance with a certain structure, a questionnaire was developed, on the basis of which a group of experts examined the technical condition of 70 electrical installations of an enterprise engaged in processing agricultural products. Based on these data, a knowledge base was formed, consisting of 42 input parameters characterizing 70 electrical installations of the agro-industrial complex of the Altai region.

4. Implementation of calculations

Based on the analysis of works in this field [6, 7], neural networks were chosen as a method for calculating the final value of fire risk, as a tool that allows finding solutions with implicit patterns and is resistant to inaccuracies in input data, which are possible due to the human factor when using a system of expert assessments. [10]

Using a neural network requires a training sample, which was formed from the knowledge base and amounted to 70 examples with 42 estimated parameters. With such characteristics, high-quality training of a neural network is difficult, and collecting additional examples involves significant time and cost costs. In this regard, it was decided to use mathematical methods to reduce the training sample without reducing the data.

The first one used was the classical correlation method. As part of its use, the correlation coefficients between the input values and the output parameter were calculated to determine the significance level of the input parameters. The results are presented in table 1.

Table 1 - Data on the correlation between the evaluation criteria and the final value of the fire risk of the electrical installation

№	Name of the parameter	Value of the correlation coefficient
x_{13}	Experience and seniority of the staff	0.020481
x_{14}	Training of the staff	0.084985
x_6	Ability to make independent decisions	0.225298
x_{28}	Passive means of electrical protection	0.254212
x_{23}	Structural (complete) failure of control panels, instrumentation, switching equipment	0.260194
x_2	Health status of the staff	0.295438
x_4	Psycho-emotional state of the staff	0.303684
x_{30}	Amount of fire damage	0.305517
x_{12}	Availability of specialized education	0.338251
x_{31}	Preventive measures level	0.347354
x_1	Physique	0.368032
x_9	Fixing technological violations	0.378906
x_{24}	Number of potentially dangerous areas for personnel	0.383887
x_{16}	Wear rate of insulating elements	0.385668
x_{29}	Fire frequency	0.395843
x_{42}	R&D	0.414495

x_{32}	Temperature differences in the working space	0.414848
x_{19}	Functional (partial) failure of technological equipment	0.419798
x_{33}	Humidity level in the working space	0.420626
x_{17}	Wear rate of structural elements	0.422469
x_{22}	Structural (complete) failure of technological equipment	0.423927
x_{41}	Introduction of new equipment	0.424367
x_{38}	The economic condition of the enterprise	0.426936
x_{37}	Compliance laws in work and industrial safety	0.431471
x_{15}	Wear rate of current conductors	0.433006
x_{27}	Active means of electrical protection	0.435697
x_3	Age of the staff	0.457453
x_{35}	Compliance with GOSTs, IEC requirements	0.476072
x_{25}	Possibility of forced or accidental stay in the danger zone	0.479392
x_{39}	Adequacy of funding	0.490119
x_{10}	Implementation of job descriptions	0.492634
x_{26}	Duration of forced stay in hazardous areas	0.494374
x_7	Workplace vigilance	0.494801
x_{36}	Compliance level with technical and sanitary standards	0.496392
x_{40}	Planning and control system	0.502933
x_5	Self-control in extreme situations	0.515955
x_{34}	Biological organisms in the working space	0.521942
x_{20}	Functional (partial) failure of control panels, instrumentation, switching equipment	0.537395
x_{11}	Quality and frequency of instruction	0.547130
x_8	Knowledge and implementation of job descriptions	0.549767
x_{21}	Structural (complete) failure of electrical network, wiring	0.563531
x_{18}	Functional (partial) failure of electrical network, wiring	0.591250

Based on the data obtained, training samples were formed for the neural network, each sample was formed by sequentially adding input parameters, starting with the most significant and ending with the less significant ones. The decisive value for the training and test samples was the assessment made by experts when assessing the fire risk of an electrical installation. The results of the training are shown in Table 2.

Table 2. Values of neural network training errors for the parameters that have the greatest correlation with the decisive value

Input parameter set	Training		Testing		Average error on training and test sets
	Average	Max	Average	Max	
x18	0,0912	0,4245	0,1132	0,2497	0,1022
x18 x21	0,0658	0,4695	0,0671	0,1543	0,0665
x18 x21 x8	0,0509	0,3783	0,0823	0,3167	0,0666
x18 x21 x8 x11	0,0330	0,2874	0,0636	0,2500	0,0483
x18 x21 x8 x11 x20	0,0351	0,2503	0,0053	0,0156	0,0202
x18 x21 x8 x11 x20 x34	0,0255	0,2776	0,0572	0,2238	0,0413

The analysis of errors on the training and test samples showed that the smallest error value when using 6 parameters, such as "Biological organisms in the space", "Functional (partial) failure of control panels, instrumentation, switching equipment", "Quality and frequency of instruction", "Knowledge and execution of job descriptions", "Structural (complete) failure of the electrical network, wiring", "Functional (partial) failure of the electrical network, wiring", 0.0202

With a further increase in the number of parameters, the error tends to increase, as can be seen from the graph shown in Figure 2.

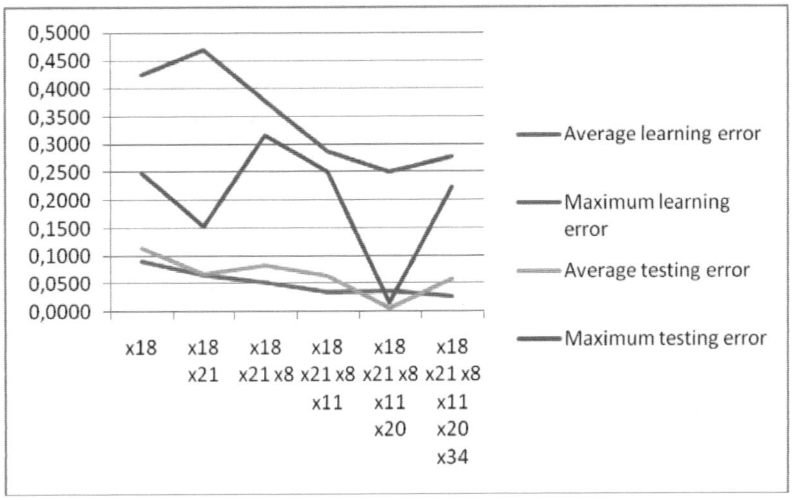

Figure 2. Graphical representation of the neural network training errorfor calculations using the correlation coefficient

However, despite the rather small error value, using this method, it must be recognized that its application is not optimal. The number of parameters on which the calculation is based is reduced to 6, and does not take into account other parameters that are also recognized by experts as important when calculating integral risk.In this regard, it was decided to use the principal component method, which allows you to take into account all the parameters.

A number of experiments were conducted to determine the optimal number of calculated principal components.

Based on the sample was calculatedthe values of the main components, then a neural network was trainedon the resulting data.

The calculation results are presented in Table 3.

Table 3. Values of neural network training errors based on the sample formed on the basis of the principal component method

Number of input main components	Training		Testing		Average error on training and test sets
	Average	Max	Average	Max	
2	0,008	0,083	0,076	0,118	0,042
3	0,006	0,099	0,059	0,115	0,032
4	0,005	0,096	0,052	0,101	0,028
5	0,002	0,039	0,065	0,092	0,033
6	0,003	0,047	0,036	0,039	0,019
7	0,003	0,066	0,084	0,105	0,044

Leaving the resulting function unchanged, the neural network was trained for each of the newly generated sets of components.

Comparison of the smallest values of the training and test errors of the correlation methods and the main components shows a slight difference.

However, the main component method allows taking into account all the initial parameters in the calculations, making it more preferable for application. Since, given the uncertainty under which the settlement are carried out, a possible false-positive triggering of the system is more preferable than skipping an event that may lead to a fire-hazardous situation.

In accordance with the tasks set in this paper, a conceptual model of the software implementation of the developed methodology has been developed (Figure 3).

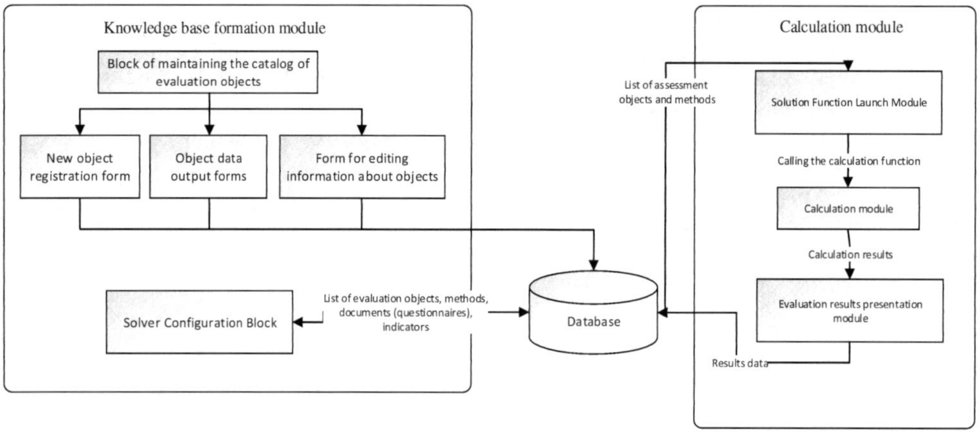

Figure 3.Conceptual scheme of software automation of the developed calculation method

The software package consists of two modules: “Knowledge base formation module” and “Calculation module”.

“Knowledge base formation module” contains block of maintaining the catalog of evaluation objects, where combines dataabout organizations and electrical installations operated on them. And the form of editing this data.The entered data about the objects are recorded in the database. This module also contains a block for configuring the solution model.

“Calculation module” consists “Solution Function Launch Module” where is List of assessment objects and methods transmitted from the database. Based on the configured solution system, the calculation is performed using the principal component method and subsequent calculation based on a trained neural network.

The result obtained is given in numerical value, on its basis it is possible to make management decisions on carrying out preventive measures.

As part of further development, a software implementation of a simulation model is planned, which would carry out calculations with changing input parameters and generate recommendations on carrying out work to reduce the level of fire hazard risk.

5. Conclusion

The developed methodology makes it possible to calculate the fire risks of the operation of electrical installations of the agricultural and industrial complex based on the concept of the man-machine system "Man-Electrical installation-Environment".

References

[1] State report "On the state of protection of the population and territories of the Russian Federation from natural and man-made emergencies in 2020" / - Moscow: EMERCOM of Russia. FSBI VNII GOChS (FC), 2021, 264 p.

[2] Shanygin, I.A. Management of technogenic risks and optimization of the safety system of electrical installations of the agro-industrial complex infrastructure: dis. candidate of Technical Sciences: 05.20.02. - Ulan-Ude, 2020. - 189 p.

[3] Kachesova, L.Y. Diagnostics of the danger of electrical installations of agricultural facilities based on a fuzzy temporal model of decision-making : dis. candidate of Technical Sciences: 05.20.02. - Barnaul, 2019. - 127 p.

[4] Eremina, T. V., Garmaev A. L. Systematic analysis of technogenic safety of electrical // Bulletin of ESSUTM 2016. № 5(62). S. 19-24.

[5] Nikol'skii, O. K. Methodology of management of technogenic risk of danger of electrical objects APK / O. K. Nikol'skii, N. P. Vorob'ev, L. V. Kulikova, E. O. Martco, D. A. Boyarkov //

Achievements of science and technology of the agro-industrial complex. - Moscow, 2017. Vol.31. - No. 3.- pp. 58-64.

[6] Eremina T.V., Shanygin I.A. Method of evaluation and optimization of fire safety of electrical installations of production facilities// Russian science in the modern world. 2019. pp. 46-48.

[7] Kalinin, A. F. Assessment and management of the integral hazard risk of electrical installations at agricultural enterprises in conditions of uncertainty: abstract of the dissertation of the Candidate of Technical Sciences. 05.20.02 / Kalinin Alexander Fedorovich.Ulan-Ude, 2015. 40 p .

[8] Kalinin, A. F. Assessment and management of integral hazard risk of electrical installations at agricultural enterprises in conditions of uncertainty: abstract of dis. ... Candidate of Technical Sciences.05.20.02 / Kalinin Alexander Fedorovich.Ulan-Ude, 2015. 40 p .

[9] Kalinin, A. F., Eremina T.V. The main directions of improving the safety of electrical installations // Energy and resource conservation - XXI century. Orel. 2017. pp. 27-30.

[10] Chupakova, A. O. Solving problems of neural network classification within the calculation of fire risks at production facilities // Informatics: problems, methods, technologies. Voronezh. 2021. - pp. 1545-1550

The development of an energy efficient electric drive for agricultural machines

T M Khalina[1], S Yu Eremochkin[1] and D V Dorokhov[1]

[1] Polzunov Altai State Technical University, 46 Lenin Avenue, Barnaul, Russia

Abstract. Agriculture is a socially significant sector of the economy. The growth of agricultural production contributes to the stable development of society. It is necessary to use new mechanisms driven by induction motors to increase agricultural productivity. Three-phase induction motors are mainly used in the electric drive of agricultural machines. At the same time, it is advisable to use a single-phase network to supply power to remote farms. In this regard, the development of a single-phase electric drive using three-phase motors becomes relevant. In this work, a study of an original semiconductor device for starting a three-phase induction motor from a single-phase network is made. The simulation model of the device created in the Matlab Simulink environment made it possible to study the electromechanical characteristics of the induction motor when operating from a single-phase network. A comparison of the characteristics of the motor during operation from a three-phase and a single-phase network is carried out. The most significant results of the work are the data obtained that the developed device can be used to start and operate a squirrel cage induction motor from a single-phase network. At the same time, the engine energy parameters change slightly.

1. Introduction

At present, agriculture in Russia has great potential for development. If in large agrarian enterprises, modern technological complexes and electrified mechanisms are already used to increase labor productivity and increase the output of finished products, then in small farms, as a rule, outdated production methods are still used.

A great economic effect is achieved when new mobile electrified machines are introduced into agricultural production: feed grinders, mechanical conveyors, electric pumps, milking machines, etc. The use of the above devices leads to an increase in labor productivity, reduces maintenance and repair costs, and improves working conditions.

Induction motors are used to drive machines, mechanisms and units. In most cases, a three-phase asynchronous electric drive is used in agricultural devices. Induction motors has the advantages: lower capital costs, higher efficiency, low maintenance costs. The disadvantage of a three-phase electric drive is the rise in the cost of the power supply system due to the need to use three-core cables. For this reason, it is often more economically justified to use a cheaper single-phase power supply system for remote farms.

Thus, an economical electric drive should combine the advantages of a three-phase and a single-phase power supply system and be devoid of their disadvantages. Therefore, the problem of choosing an effective method for starting a three-phase induction motor from a single-phase network arises.

2. Development of a simulation model of a three-phase induction motor

The «Semiconductor device for starting a three-phase induction motor from a single-phase network» was developed at the Polzunov Altai State Technical University by the staff and students of the department «Electrical Engineering and Automated Electric Drive» [1]. The schematic diagram of the device is shown in Figure 1.

There are similar devices that also allow starting three-phase induction motors from a single-phase network, made on the basis of inductance and a large paper capacitor [2]. They have a number of disadvantages: low reliability, unpredictability rotation direction of the engine, low energy performance. All these disadvantages are eliminated in the semiconductor starting device due to the original engine control system, made on the basis of two power semiconductor switches. IGBT, MOSFET transistors and others can be used as semiconductor switches.

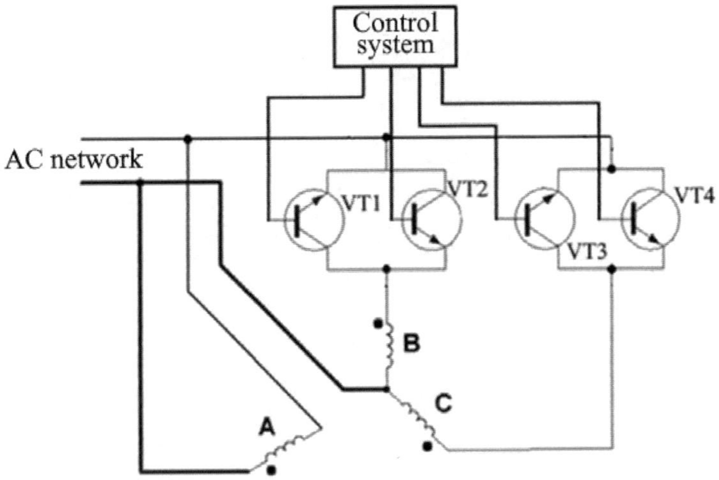

Figure 1. Schematic diagram of a semiconductor device for starting squirrel cage induction motor from a single-phase network.

To test the performance of such devices, it is necessary to create prototypes. At the initial stages of development, this is economically unprofitable, and in conditions of limited funding, it is often not possible. Therefore, scientists and researchers are increasingly resorting to computer simulation tools [3-13]. Their features include the relative cheapness of research, the ability to simulate the behavior of the system in time. One of these tools is the interactive graphical environment for simulation Matlab Simulink. This environment allows us to simulate electrical machines, and also has a graphical presentation of the simulation results [14-17].

To carry out a computer experiment, a mathematical model of the motor is required. Simulink standard blocks cannot be used for the correct experiment, since the stator windings are fed asymmetrically with vector-algorithmic motor control. In this regard, an original mathematical model of an induction motor was developed. This model allows calculating static and dynamic modes with an asymmetric power supply [18]. On the basis of a mathematical model, a simulation model was built in the Matlab Simulink environment. According to mathematical model the electromechanical characteristics of a three-phase induction motor with a semiconductor starting device were calculated.

The simulation model was made on the basis of the parameters of a real motor (AIS56V4U3). This squirrel-cage induction motor has the following nominal characteristics: power, $P_r = 90$ W; voltage, $U_r = 380\ V$; current, $I_r = 0.42\ A$; the moment developed by the motor, $T_r = 0.64$ Nm; rotation frequency, $n = 1350\ rpm$; efficiency value, $\eta_r = 55\%$; the value of the power factor, $cos\varphi = 0.6$, as well as the multiplicity of the starting torque, $k_s = T_s/T_r = 2.1$; multiplicity of the maximum torque, $\lambda = T_{max}/T_r = 2.2$; multiplicity of the minimum torque, $k_{min} = T_{min}/T_r = 1.8$. The stator windings of the induction motor are star-connected.

3. Operation investigation of an electric drive powered from a three-phase network

At the first step, the electromechanical characteristics of the motor powered from a three-phase network were calculated. The obtained results were compared with the semiconductor starting device

characteristics. Figure 2 shows a simulation model of the direct start of an induction motor, created in the Simulink environment.
The electric drive in the simulation model is powered from ideal AC voltage sources that simulate a three-phase network from the standard Electrical Sources library. The characteristics of the induction motor were recorded using measuring units «Scope».

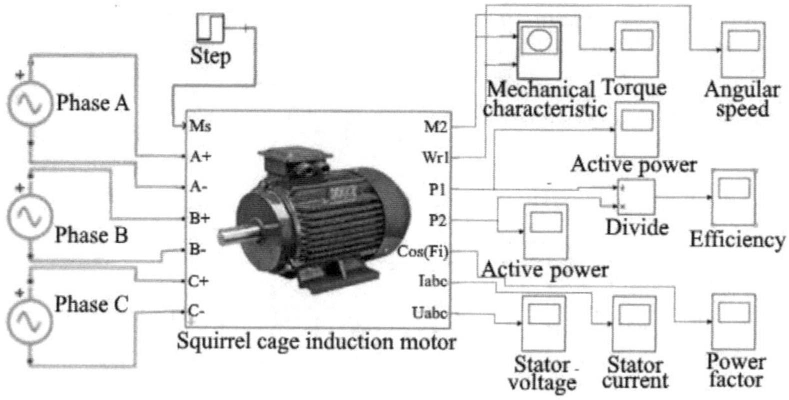

Figure 2. Model of squirrel cage induction motor with direct start in the Simulink environment.

The simulation results are presented in Figures 3 - 8.

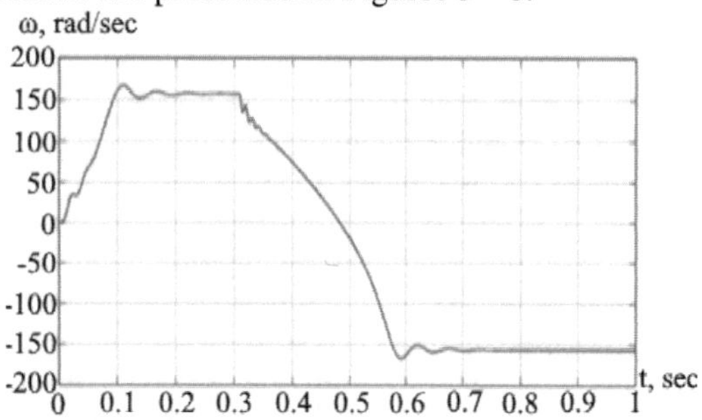

Figure 3. Oscillogram of angular velocity change in start-up mode, anti-inclusion braking and reverse.

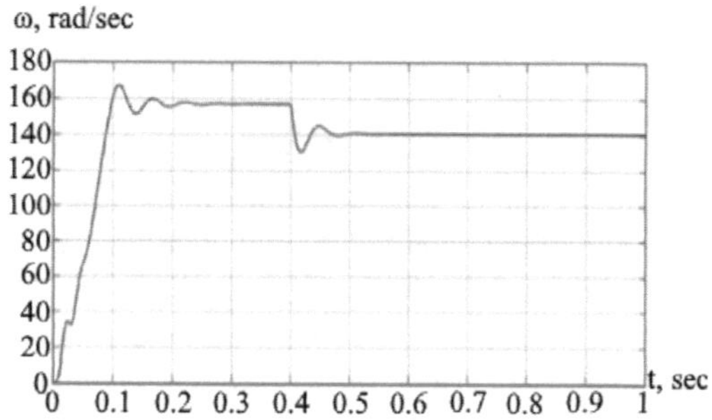

Figure 4. Oscillogram of angular velocity change in start-up and operation under load.

Figure 3 shows that the steady state of the angular speed of the motor shaft at start-up comes in 0.2 sec. Opposition braking starts at 0.3 sec and ends at 0.48 sec (duration 0.18 sec). In the steady state, the angular velocity of the motor shaft during reverse occurs at the simulation time of 0.68 sec, respectively, the time until the steady state at start and reverse is the same (equal to 0.2 sec).
Figure 4 shows how the change in the angular speed of the induction motor shaft occurs at start-up at the time from the start of the simulation (0 sec) to the instantaneous change in the nominal load torque (0.64 Nm) after 0.4 sec on the shaft. It should be noted that after connecting the load to the steady-state value, the angular velocity comes in 0.1 sec.
Figure 5 shows how the torque on the motor shaft changes during the simulation time. In the period from the start of the simulation to the start of the anti-rotation braking, the torque fluctuates from 1.8 Nm to -0.5 Nm and tends to a steady-state value of 0 Nm. The torque during opposing braking, at a time from 0.3 sec to 0.48 sec, ranges from -3.3 Nm to 1.8 Nm and tends to a steady-state value of -1 Nm. The torque during reverse, at a time from 0.48 syc until the end of the simulation time, ranges from -1.5 Nm to 0.5 Nm and tends to a steady-state value of 0 Nm.

Figure 6 shows that at start-up, the torque on the motor shaft, at times from the start of the simulation (0 sec) and up to 0.4 sec, fluctuates from 1.8 Nm to -0.5 Nm and tends to a steady-state value of 0 Nm. With an instantaneous change in the nominal load torque (0.64 Nm) after 0.4 sec, the torque fluctuates from 0 Nm to 1 Nm and tends to a steady-state value of 0.64 Nm.

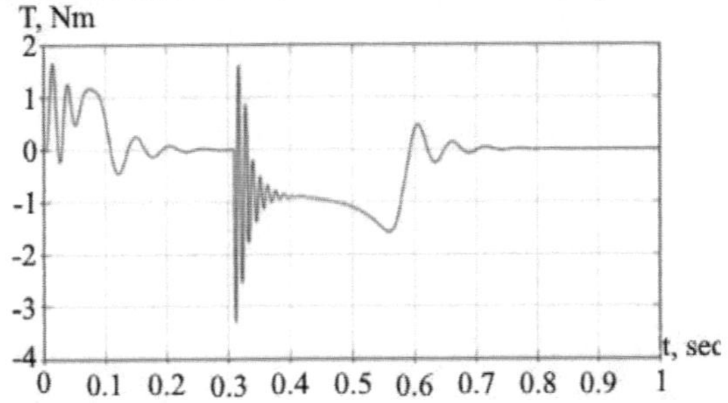

Figure 5. The oscillogram of the change in torque on the shaft in start-up mode, anti-inclusion braking and reverse.

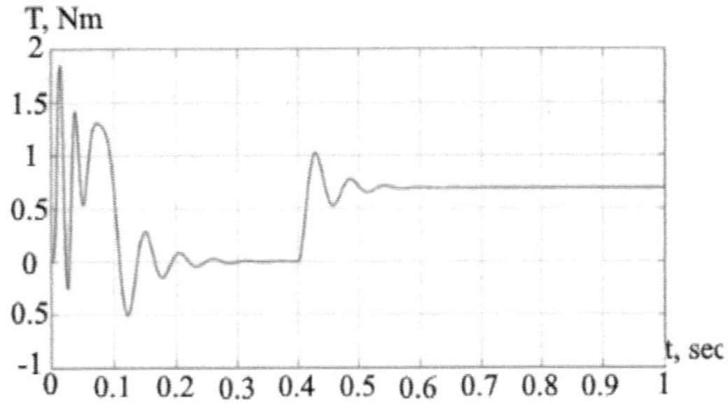

Figure 6. The oscillogram of the change in torque on the shaft in the mode of start-up and operation under load.

Figure 7 shows that the efficiency of the electric motor at rated load is $\eta \approx 55\%$, the power factor is $cos\varphi \approx 0.6$. These values are close to the motor manufacturer's ratings.
Figure 8 shows the value of the starting torque of the induction motor, which is $T_s = 1.3$ Nm, while the value of the critical torque is $T_{cr} = 1.35$ Nm.
Based on the obtained data, it can be concluded that the obtained characteristics of the induction motor are close to the nominal ones.

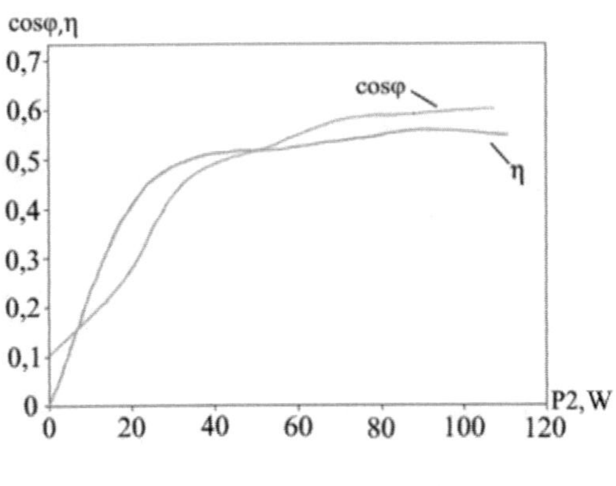

Figure 7. Three-phase mains motor performance.

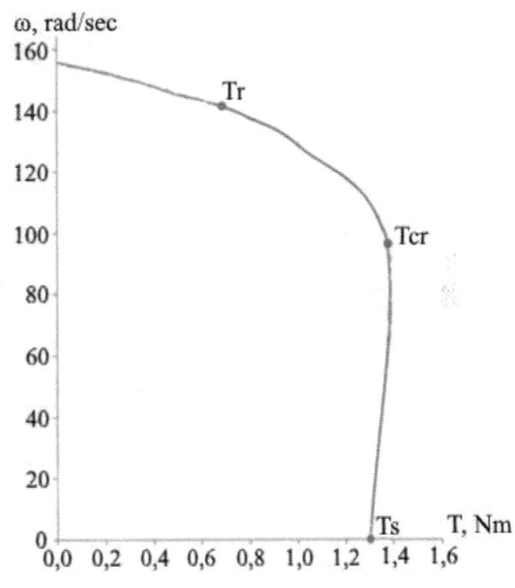

Figure 8. Mechanical characteristics of the motor with three-phase power.

4. Investigation of the operation of an electric drive with a vector-algorithmic type converter using the developed simulation model

Figure 9 shows a model of an electric drive with the motor and a semiconductor starting device, created in the Simulink simulation environment.

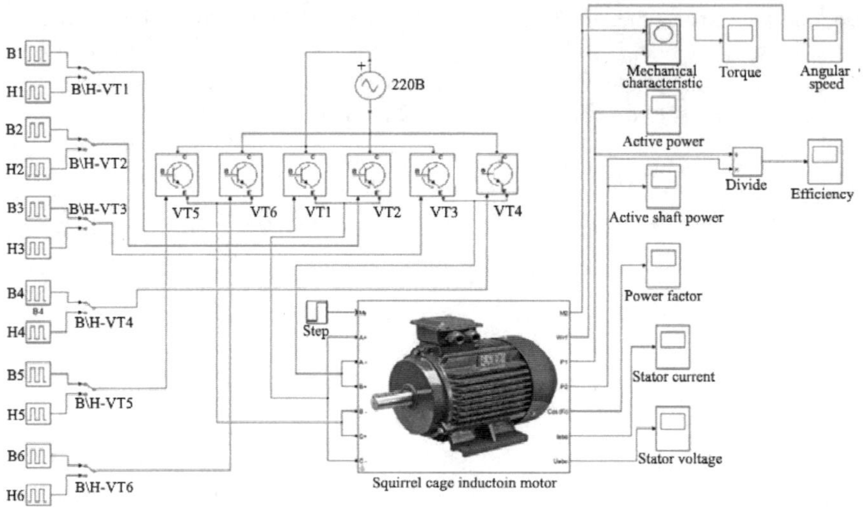

Figure 9. Model of an electric drive with a semiconductor starting device in a Simulink environment.

In this model, as in the previous one, the transistor control system is simulated using pulse generators. They are easy to configure for issuing control signals with the desired frequency and amplitude. To simulate the operation of the device according to the clockwise field rotation algorithm, the «B / H» switches for all transistors are moved to the position to the upper signal generator «B», to simulate the operation of the device counterclockwise, the switches are moved to the position to the signal generator «H».
Models of «n-p-n» bipolar transistors are taken from the standard library «Power Electronics», which is included in the library for modeling electrical devices and energy systems «SimPowerSystems». Transistor models are accepted as ideal semiconductor switches that do not alter the operation of the drive system.

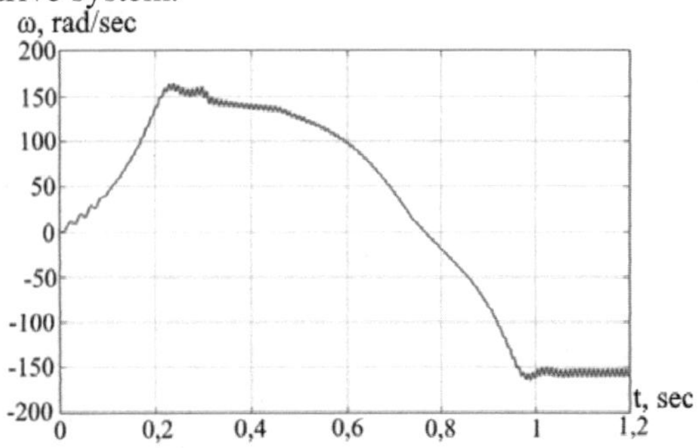

Figure 10. Oscillogram of angular velocity change in start-up mode, anti-inclusion braking and reverse.

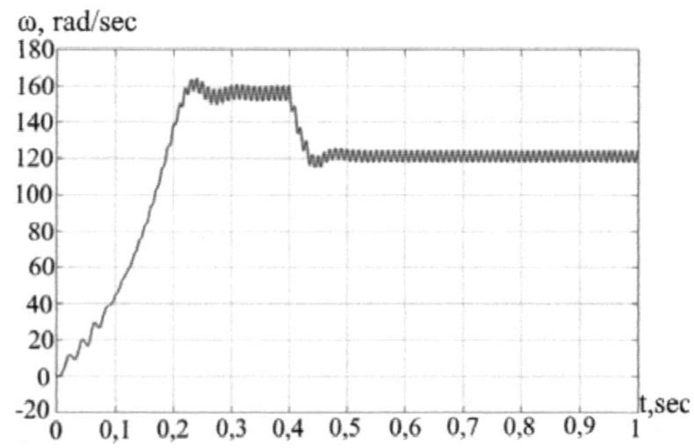

Figure 11. Oscillogram of angular velocity change in start-up and operation under load.

Figure 10 shows that the angular speed of the motor shaft at start-up reaches a steady state in 0.3 sec. Opposition braking starts at 0.3 sec and ends at 0.78 sec (duration 0.48 sec). The steady state angular speed of the motor shaft during reverse reaches 1.08 sec at the time of simulation, respectively, the time until the steady state at start and reverse is the same and is 0.3 sec. In comparison with direct starting, there is an increased acceleration time by 0.1 sec and an increase in the duration of the anti-switching braking time by 0.3 sec.
Figure 11 shows how the change in the angular speed of the induction motor shaft occurs at start-up at the time from the beginning of the simulation (0 sec) to the instantaneous change in the nominal load

torque (0.64 Nm) after 0.4 sec on the shaft. It should be noted that after connecting the load, the angular velocity reaches a steady-state value (123 rad/sec) in 0.12 sec.

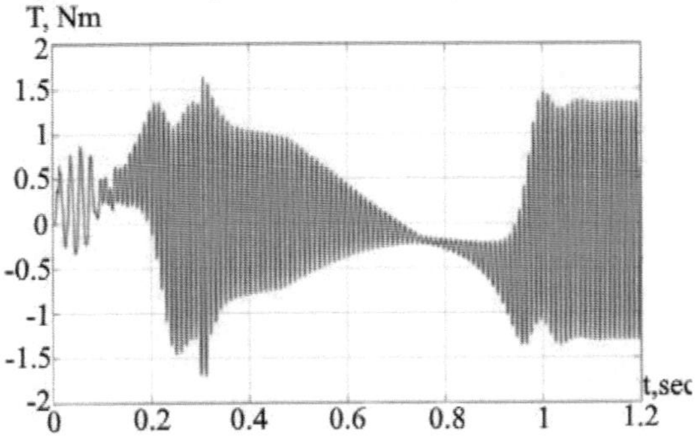

Figure 12. The oscillogram of the change in torque on the shaft in start-up mode, anti-inclusion braking and reverse.

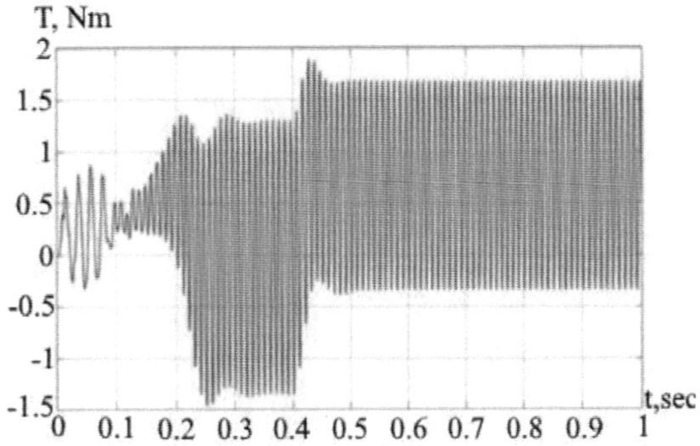

Figure 13. The oscillogram of the change in torque on the shaft in the mode of start-up and operation under load.

Figure 12 shows how the torque on the motor shaft changes during the simulation time. In the period from the start of the simulation to the start of the anti-rotation braking, the torque fluctuates from 1.3 Nm to -1.4 Nm. The torque during opposing braking, in the time interval from 0.3 sec to 0.78 sec, ranges from -1.6 Nm to 1.6 Nm and tends to an average steady-state value of -0.3 Nm. The torque during reverse, in the period from 0.78 sec to the end of the simulation time, ranges from 1.4 Nm to -1.3 Nm.

Figure 13 shows that at start-up, the torque on the motor shaft at times from the start of simulation (0 sec) and up to 0.4 sec, ranges from 1.3 Nm to -1.4 Nm. With an instant change in the rated load torque (0.64 Nm) after 0.4 sec, the torque fluctuates from 1.7 Nm to -0.4 Nm.

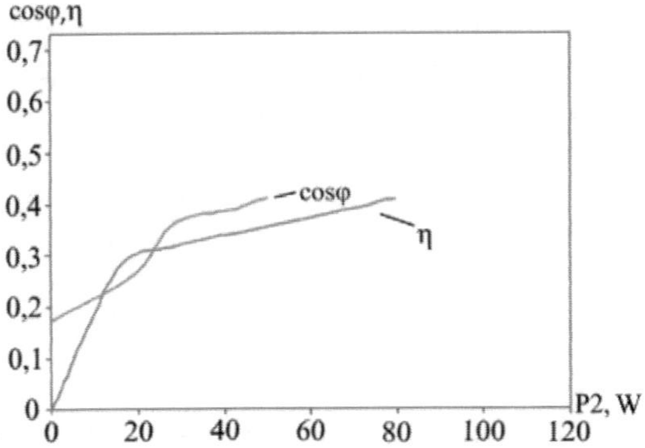

Figure 14. Semiconductor starter motor performance.

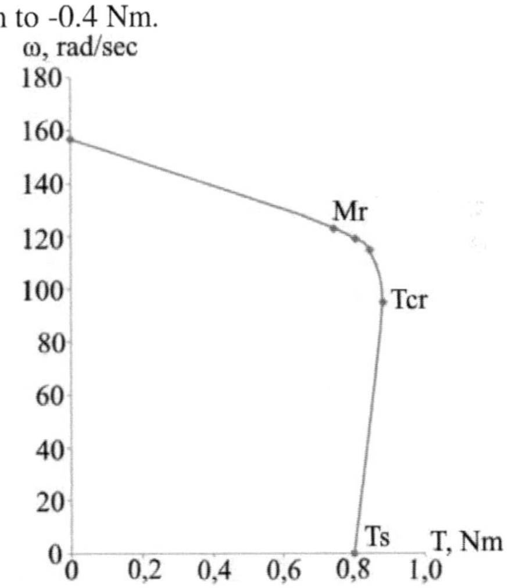

Figure 15. Static mechanical characteristic of squirrel cage induction motor with a semiconductor starting device.

Figure 14 shows that the power of the electric motor when working with a semiconductor starting device is slightly less than the value of the rated power. This is due to the fact that the stator windings are not supplied simultaneously. In this case, the efficiency reaches the value $\eta \approx 40\%$, the power factor $cos\varphi \approx 0.42$.

Figure 15 shows that the starting torque of the induction motor is $T_s \approx 0.8$ Nm, the critical moment is $T_{cr} \approx 0.85$ Nm.

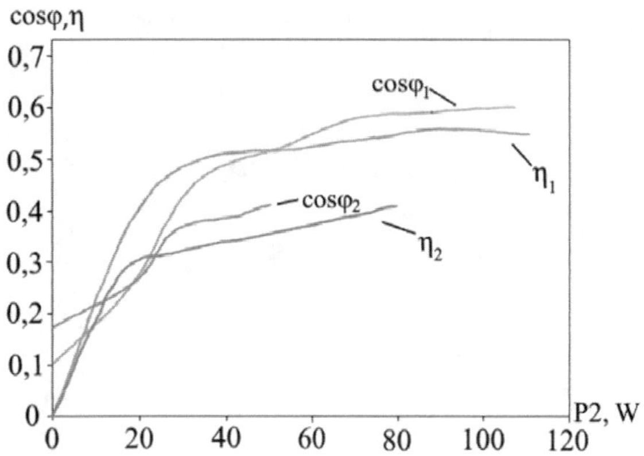

Figure 16. Combined engine performance $cos\varphi_1$ – power factor with three-phase power supply; η_1 – efficiency at three-phase power supply; $cos\varphi_2$ – power factor when working with a semiconductor starting device; η_2 – efficiency when working with a semiconductor starting device.

Figure 16 shows that the values of the power factor and the efficiency of the induction motor in both cases practically coincide at power values up to 50 W.

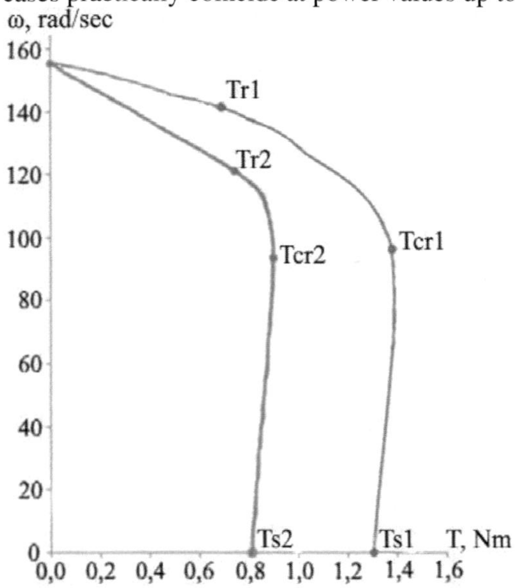

Figure 17. Combined mechanical characteristics of an induction motor T_{r_1} – rated torque at three-phase power supply; T_{cr_1} – critical moment with three-phase power supply; T_{s_1} – starting torque with three-phase power supply; T_{r_2} – rated torque when working with a semiconductor starting device; T_{cr_2} – critical moment when working with a semiconductor starting device; T_{s_2} – starting torque when working with a semiconductor starting device.

Figure 17 shows that the value of the starting torque of an induction motor with a semiconductor starting device, which is equal to $T_s \approx 0.8$ Nm, which is 54% of the starting torque when powered from a three-phase network. The value of the critical torque of the induction motor is equal to $T_{cr} \approx 0.85$ Nm, which is 56% of the critical torque when powered from a three-phase network.

5. Conclusion

Thus, as a result of the study the characteristics of the three-phase induction motor with a semiconductor starting device operating from a single-phase network, in the Matlab Simulink simulation environment were modelled. It was found that the starting torque of the engine is $T_s \approx 0.8$ Nm, the critical torque is $T_{cr} \approx 0.85$ Nm, the efficiency reaches the value $\eta \approx 40\%$, the power factor is equal to $cos\varphi \approx 0.42$. At the rated torque, the speed of the induction motor when operating on an artificial characteristic is 12% less than the speed of the induction motor on a natural characteristic, which slightly reduces the performance of the electric drive.

Based on the foregoing, it can be concluded that the use of the developed electric drive with a semiconductor starting device in the absence of a three-phase power supply is fully justified.

Acknowledgments

The work, based on the results of which the article was completed, was carried out under a grant from the President of the Russian Federation for state support of young Russian scientists - candidates of sciences on the topic "Development and research of an energy-efficient reversible semiconductor switch, driven by a network, for single-phase electric drives of agricultural and urbanized machines of low power". Grant No. MK-4018.2021.4. Agreement No. 075-15-2021-087 dated 04.16.2021.

References

[1] Stalnaya M I, Eremochkin S Y, Dorokhov D V 2019 Semiconductor device for starting a three-phase asynchronous electric motor from a single-phase network *Patent RF* no. 192777
[2] Glazenko T A and Khrisanov V I 1983 *Semiconductor systems of low-power impulse asynchronous electric drive* (Leningrad: Energoatomizdat) p 176
[3] Huan-Liang Tsai and Phuong Truong Le 2016 Self-sufficient energy recycling of light emitter diode/thermoelectric generator module for its active-cooling application *Energy Conversion and Management* **118** 170-8
[4] Ouanjli N El, Motahhir S, Derouich A, Ghzizal A El, Chebabhi A, Taoussi M 2019 Improved DTC strategy of doubly fed induction motor using fuzzy logic controller *Energy Reports* **5** 271-9
[5] Abdin Z, Webb CJ, Mac E and Gray A 2018 One-dimensional metal-hydride tank model and simulation in Matlab–Simulink *International Journal of Hydrogen Energy* **43**(10) 5048-67
[6] Sung-ho Hur 2018 Modelling and control of a wind turbine and farm *Energy* **156** 360-70
[7] El-Kharashi E, Massoud J G, Al-Ahmar M A 2019 The impact of the unbalance in both the voltage and the frequency on the performance of single and cascaded induction motors *Energy* **181** 561-75
[8] Zhifu W, Jun F, Zhijian S and Qiang S 2017 Study on Speed Sensor-less Vector Control of Induction Motors Based on AMEsim-Matlab/Simulink Simulation *Energy Procedia* **105** 2378-83
[9] Abdin Z, Webb C J, Mac E and Gray A 2016 PEM fuel cell model and simulation in Matlab–Simulink based on physical parameters *Energy* **116**(1) 1131-44
[10] Musio F, Tacchi F, Omati L, Stampino PG, Dotelli G, Limonta S, Brivio D and Grassini P 2011 PEMFC system simulation in MATLAB-Simulink environment *International Journal of Hydrogen Energy* **36**(13) 8045-52
[11] Ilimbetov R Yu, Popov V V and Vozmilov A G 2015 Comparative Analysis of "NGTU – Electro" Electric Car Movement Processes Modeling in MATLAB Simulink and AVL Cruise Software *Procedia Engineering* **129** 879-85
[12] Asbjörnsson G, Hulthén E and Evertsson M 2013 Modelling and simulation of dynamic crushing plant behavior with MATLAB/Simulink *Minerals Engineering* **43** 112-120
[13] Bhola M, Kumar N and Ghoshal S K 2018 Reducing fuel consumption of Front End Loader using regenerative hydro-static drive configuration-an experimental study *Energy* **162** 158-170
[14] Xiaolin Tang, Dejiu Zhang, Teng Liu, Amir Khajepour, HaishengYu and Hong Wang 2019 Research on the energy control of a dual-motor hybrid vehicle during engine start-stop process *Energy* **166** 1181-93
[15] Guoheng Wu, Junhong Yang, Jianzhong Shang and Delei Fang 2020 A rotary fluid power converter for improving energy efficiency of hydraulic system with variable load *Energy* **195** 116-957
[16] Eyhab El-Kharashi and Maher El-Dessouki 2014 Coupling induction motors to improve the energy conversion process during balanced and unbalanced operation *Energy* **65** 511-16
[17] Ahmed Abrarм, Yelamali Palaksh and Udayakumar R 2020 Modelling and simulation of hybrid

technology in vehicles *Energy Reports* **6** 589-94
[18] Stal'naya M I, Eremochkin S Y 2015 Simulation of the electromechanical characteristics of a three-phase electric motor with a vector-algorithmic type converter in MATLAB SIMULINK.
Elektroprivody peremennogo toka: Trudy shestnadcatoj Mezhdunarodnoj nauchno-tekhnicheskoj konferencii. Ekaterinburg 145-8

STUDY OF THE THERMAL PROCESSES DYNAMICS IN THE FEEDSTUFF DISINFECTION BY ELECTRIC CONTACT HEATING

T M Khalina[1], M V Khalin[1], M V Dorozhkin[1]

[1]Polzunov Altai State Technical University (ASTU), 656038, pr. Lenina, 46, Barnaul, Russia

E-mail: temf@yandex.ru

Abstract. The article is dedicated to the study of the heat transfer processes that occur in the feedstuff disinfection chamber that relies upon electric contact heating. The mechanism of the temperature gradient appearance, which is the main cause of the heat losses has been investigated. The basic equations of heat conduction are considered. A method is proposed for determining the key parameters of the heat transfer process. A functional diagram of the experimental setup with a description of the operation of individual units is presented. The dependence for the transient operating mode of the unit on the growth of heat losses has been established. Thermal images of different shapes of the unit dielectric chambers have been provided as well as temperature field distribution through the chamber wall.

Keywords: electric contact heating, feedstuff disinfection, heat losses, heat transmission coefficient, heat exchange.

On the current markets, environmentally friendly and less resource-intensive technologies are usually more preferable. As a rule, the efficiency of such systems is associated with thermal processes, and primarily heat loss. The study of the mechanism of heat losses, in particular, the analysis of the factors that have the greatest influence on their level, is one of the main tasks as regards the increase in energy efficiency [1]. The above is more true for the processes that provide changes in the thermophysical characteristics of the feedstuff, for example, the technology of electric contact (EC) disinfection of wet feed mixtures used in animal husbandry [2].

The EC disinfection belongs to the thermal feedstuff processing methods. The feedstuff disinfection effect is achieved through the exposure of pathogenic organisms to the temperature (60…80 °C) that is fatal for their life [3, 4]. The EC technology is based upon the volumetric heating of the feed mixture by the heat generated as a result of the current being passed through the feedstuff disinfected [2]. Accordingto Joule-Lenz's law, the amount of heat generated in an elementary volumecan be described by the following equation [5, 6]:

$$dQ = I^2 R dt, \tag{1}$$

where$I$ – RMS current, A; R–resistance of a conductive medium, Ohm.

Heat generation in the feed mixture leads to an increase in its internal temperature. As a result of the fact that the container in which the heating occurs has contact with a colder carrier, a temperature gradient arises at the boundary of the feed mixture - the chamber wall - air, due to which part of the heat is spent through heat exchange processes for parasitic heating of the environment. The greater the temperature gradient between hot and cold objects the higher the heat loss, provided that the remaining parameters of the system are constant [7].

The main task in determining the amount of heat transferred through an arbitrary surface is to find the characteristic points of the temperature field inside the object under consideration

[1, 7-10]. As a rule these points are the temperatures of the hot (feed mixture) and cold (air) coolants, as well as the temperatures of the inner and outer walls of the partition dividing them.

An EC unit operating process may be divided into two stages:

- continuous heating of the feed mixture with the achievement of the temperature required (transient mode);

- keeping the feed mixture under a constant temperature during time established under the process (steady state).

Each of the modes suggests its own approach to the establishing thermal parameters dynamics.

Procedure for Determination of Heat Losses

For the steady-state mode and the boundary conditions of the third kind (Figure 1) the amount of heat transmitted by the heated mixture to the wall by means of heat exchangeis determined based on the Newton-Richmann equation [1]:

$$Q_1 = \alpha_1 F (t_m - t'_w), \tag{2}$$

whereα_1 – coefficientofheattransmissionbythehotmediumwithatemperatureof$t_m$tothewallsurface, W/(m^2·°C); F– flat wall surface area, m^2; t'_w– surface temperature of the wall bordering with the hot heat transfer agent, °C.

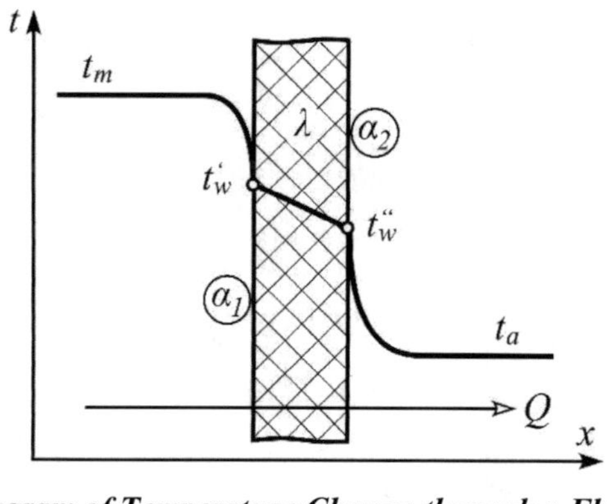

Figure 1 – Diagram of Temperature Change through a Flat Wall

Depending on the heat exchange conditions, calculation of the heat transmission coefficient α_1 can include all its types, however, as applied to the EC heating technology, there is no temperature gradient as a result of almost uniform internal heating throughout the entire volume of the feed mixture [2, 5]. Due to the insignificant movement of the liquid in the mixture under such conditions, heat exchange is by convection replaced by thermal conductivity.

Heat is transferred through a flat wall by thermal conductivity and can be described by the following equation [1]:

$$Q_2 = \frac{\lambda}{\delta} F (t'_w - t''_w), \tag{3}$$

where λ – wall material heat transmission coefficient, W/(m·°C); δ– wall thickness, m; t'_w, t''_w– temperature of the internal and external wall surfaces, °C.

Air stream from the outer wall surface to the cold air with a temperature of t_a is calculated similarly to the calculation of heat for the hot mixture - internal container wall surface region [1]:

$$Q_3 = \alpha_2 F\,(t''_w - t_a), \tag{4}$$

where α_2 – coefficient of heat transmission from the wall to the air, W/(m^2·°C).

For the calculation of heat transfer coefficient α_2, in addition to convective heat transfer between laminar air jets rising along the heated wall, radiation (radiant) component of heat transfer shall also be accounted for.

Since the values of heat streams in equations (2), (3), (4) are equal with one another: $Q_1 = Q_2 = Q_3 = Q$, because they describe different stages of the same heat transmission process;summing all resulting equations gives the following heat transmission expression [1]:

$$Q = kF\Delta t, \tag{5}$$

where: $\Delta t = t_1 - t_2$ - difference between the temperatures of the heated feed mixture and environment, °C;

$k = \left(\dfrac{1}{\alpha_1} + \dfrac{\delta}{\lambda} + \dfrac{1}{\alpha_2}\right)^{-1}$ - complex heat transmission coefficient, W/(m^2·°C).

Calculation of heat losses should be started by determining the heat transmission coefficient α_2 from the outer heated wall of the chamber to air, since in this case the measurement of the surface temperature outside the chamber by instrumental methods turns out to be much easier and more accurate.

Heat transmission coefficient α_2 will be determined as a sum of the heat transmission coefficients of convection and radiance [1]:

$$\alpha_2 = \alpha_k + \alpha_r. \tag{6}$$

Convective heat transfer depends on a large number of factors and is rather a complex process, which is a function of the physical parameters of the heat transfer agent, the speed of its movement, the shape and size of the body, therefore, the coefficient is often calculated using similarity numbers and similarity equations as follows [1, 10]:

$$\alpha_k = Nu\frac{\lambda}{l}, \tag{7}$$

where Nu – Nusselt number.

Nusselt number can be derived by calculating a general similarity equation for convective heat transfer in laminar flow:

$$Nu = 0{,}017\,\mathrm{Re}_a^{0,33}\,\mathrm{Pr}_a^{0,43}\,Gr_a^{0,1}\,(\mathrm{Pr}_a/\mathrm{Pr}_w)^{0,25}, \tag{8}$$

where Re, Pr, Gr – Reynolds, Prandtl, Grash of similarity numbers, respectively.

Heat transfer by radiation under conditions of thermodynamic equilibrium is determined on the basis of the Stefan-Boltzmann law [10, 11]:

$$\alpha_r = \varepsilon\sigma_0\frac{(t''_w + 273{,}15)^4 - (t_a + 273{,}15)^4}{t''_w - t_a}, \tag{9}$$

where $\sigma_0 = 5{,}67\cdot10^{-8}$W/(m^2·°K^4) – Stefan-Boltzmann radiation constant; ε - degree of surface blackness.

Substituting the values of the coefficients from expressions (7) and (9) into (6), we obtain the total heat transfer coefficient.

From equation (4) with a known heat transmission coefficient, as well as known temperatures of the adjoining media, it is possible to determine the magnitude of the heat rate, which will correspond to heat loss.

The calculation of the heat transmission coefficient α_1 of the transition from the heated feed mixture to the inner wall of the container is carried out using the stationary flow method based on the Newton-Richmann equation (2):

$$\alpha_1 = \frac{Q}{F\left(t_1 - t'_w\right)} = \frac{q}{t_1 - t'_w}. \quad (10)$$

The unknown temperature of the inner surface of the wall can be determined from the expression for the temperature head Δt by solving the equation with respect to the desired value:

$$\Delta t = \frac{\delta}{\lambda} q, \quad (11)$$

where$q$ – specific heat rate.

Description of the Experimental Setup

To study the thermodynamic processes occurring during the thermal inactivation of microflora, an experimental setup was developed, the functional diagram of which is shown in Figure 2.

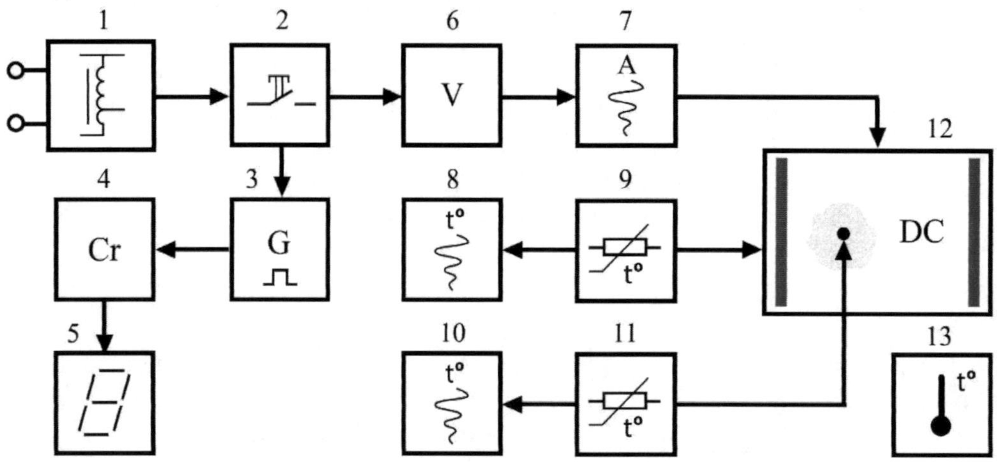

Figure 2 – Functional Diagram of the Experimental Setup

The key component of the setup is a rectangular dielectric chamber (DC) 12, made of polymethyl methacrylate (plexiglass) with a wall thickness of 5 mm and a side size of 0.02x0.12x0.23 m. Two 2mm-thick fixed steel electrodes 0.02x0.14 m in size are placed on the narrow opposite side walls inside the chamber having special connectors in the upper part for connecting power cables. The configuration of the chamber is selected in such a way as to provide the largest contact area with the environment with a sufficiently small volume of the feed mixture.

The installation is powered from a 220 V AC mains supplied to the input of laboratory autotransformer 1, at the output of which a serially connected push-button switch 2 with position fixation is installed to close the circuit. The values of the supplied voltage from the output of the autotransformer are determined by the readings of voltmeter 6, and ammeter 7 connected to the logger records the current consumed by the unit during operation [2].

To control the temperature of the outer surface of the dielectric chamber and the temperature of the feed mixture, two calibrated temperature sensors 9 and 11 based on chromel-copel thermocouples are used, connected to thermometers 8 and 10, which have the data recording feature. Digital thermometer 13 is used to determine the ambient temperature.

The duration of feed mixture heating, as well as the time to maintain the required temperature for the second stage of the experiment, are controlled by a digital stopwatch, which consists of a generator of second pulses 3 started when the switch button 2 is pressed, a pulse counter 4 and a display unit 5, necessary to display the current operating time of the unit (in seconds).

A calibrated thermal imager was used in the experiment for additional visual control of the dynamics of thermal processes, as well as determination of possible areas of inhomogeneous heating of the DC surface.

Experiment Conditions

The feed mixture, the basis of which is a complete loose compound feed used for fattening pigs, with a water mass fractionof 60% and a temperature equal to the ambient temperature of 23 °C, is loaded into the interelectrode space of the unit chamber. To prevent the feed mixture from going beyond the boundaries of the DC during heating, the volume of which can significantly increase as a result of moisture absorption by the feed components, as well as the thermal expansion of the liquid itself, an incomplete filling is provided with a volume of $4.6 \cdot 10^{-4}$ m^3 (equivalent weight - 0,41 kg).

Thermocouple 11 is placed in the inner cavity of the chamber to a depth equal to half the height of the DC, and at an equal distance between the nearest walls of the chamber. Electronic thermometer 13 is installed next to the DC of the experimental setup at a distance of up to 0.3 m.

In order to exclude the intensification of heat exchange due to the movement of air flows and, consequently, increased heat losses into the environment, the experiment was carried out in an isolated room with a uniformly distributed temperature throughout the volume.

By adjusting autotransformer 1, the supply voltage of the unit was selected based on the process conditions and amounted to 150 V for the given chamber configuration [2].

Temperature control in a steady-state mode was carried out by changing the supply voltage during the time required for complete heating of the chamber wall and reaching the steady state temperature. At this stage of the experiment, temperature control was carried out according to the readings of digital thermometer 8.

Experiment Results Analysis

According to the above procedure for determining heat losses, the experiment included two stages:

- measurement of the parameters of the installation in a non-stationary mode, when the temperature inside the DC grew continuously throughout the experiment;
- measurement of the parameters while maintaining the required temperature of the feedstuff (steady state mode).

For thetransient mode, heat transfer through the wall can be determined based on the graphs of changes in the temperature of the feed mixture t_m and the temperature t''_w of the outer wall of the DC (Figure 3).

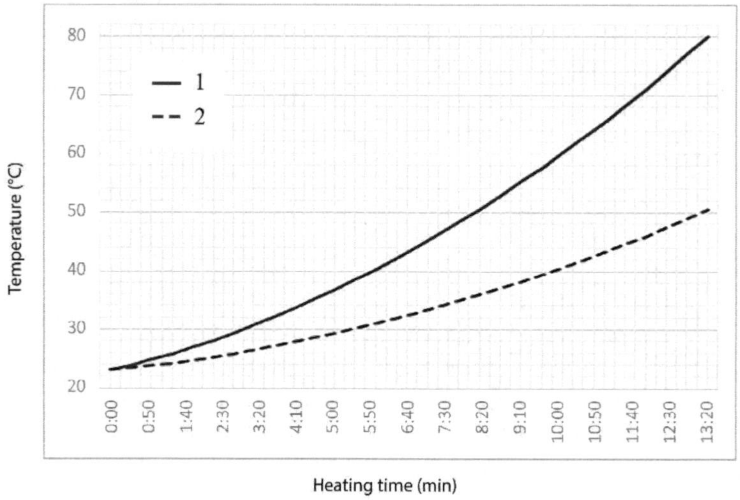

Figure 3 – Temperature-heating time relation diagram: 1 – feed mixture temperature t_m; 2 – temperature t''_w of the outer wall of the DC

The experiment has shown that as the temperature rises exponentially inside the chamber due to heating of the mixture (curve 1) the temperature head also increases (curves 1 and 2).

Due to the thermal inertia of the system resulting in rather a smooth change in the test sample temperature these temperature changes can be neglected for short time intervals, and the averaged values of temperatures for the current time interval can be used for calculations:

$$t_{avg} = \frac{1}{\tau_2 - \tau_1} \cdot \int_{\tau_1}^{\tau_2} t(\tau) d\tau, \quad (12)$$

where τ_1, τ_2 - time interval start and end points, respectively.

Then, approximate values of heat losses through the chamber walls can be obtained through applying the technique proposed for calculating the stationary flow for each sufficiently small time interval.

Figure 4 shows a graph of changes in the heat flux (curve 2), increased for clarity by 10 times. Curve 1, obtained experimentally, shows the amount of heat generation inside the entire volume.

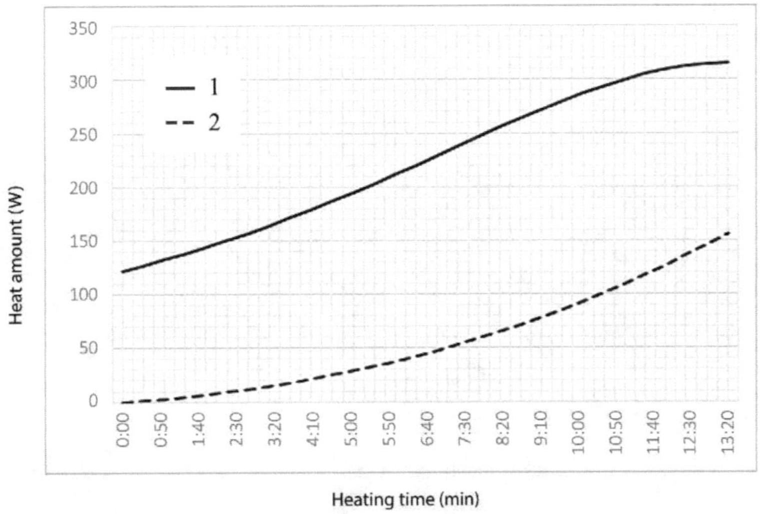

Figure4 – Heat flow-heating time ratio diagram: 1- heat Q generated in the mixture; 2 – heat losses 10· Q_l through the outer walls of the DC

Based on the data obtained (Table 1) the average heat loss during heating at any time has been found not to exceed 5% of the heat generated inside the feedstuff.

Table 1- Calculated values of heat losses during feed mixture heating

Heat generation Q, W	102	106.5	109	114	118.5	124.5	129	133.5	138	141.8
Heat losses Q_i, W	0.004	0.134	0.286	0.464	0.664	0.89	1.138	1.414	1.71	2.036
$(Q_i/Q)100$, %	0.004	0.126	0.261	0.407	0.56	0.715	0.882	1.059	1.239	1.436
Heat generation Q, W	147	149.7	154.5	159	164.7	169.5	176.2	180	183.7	189.3
Heat losses Q_i, W	2.388	2.766	3.17	3.6	4.064	4.552	5.072	5.622	6.206	6.82
$(Q_i/Q)100$, %	1.624	1.848	2.052	2.264	2.467	2.685	2.878	3.123	3.377	3.603
Heat generation Q, W	193.5	196.8	202.5	206.2	210	211.8	213.7	216	217.5	218.7
Heat losses Q_i, W	7.462	8.144	8.856	9.604	10.38	11.2	12.05	12.95	13.88	14.85
$(Q_i/Q)100$, %	3.856	4.138	4.373	4.656	4.945	5.288	5.638	5.993	6.379	6.788

The higher percentage of heat loss observed at the end of the heating (starting from 10 minutes) to the temperature required for the process (Figure 4, curve 1) is explained by the significant thermal inertia of the system: the feed mixture and the chamber wall in question, which, continues to warm up for some time (curve 2) after a decrease in the level of heat generation in the volume.

At the stage of the regulation process, with a steady state t_m = const, no significant change in the value of heat transfer from a heated wall with a temperature of 50.4 °C to an environment with a temperature of 23 °C was observed remaining at 14.86 ± 0.02 W, which indicates insignificant heat loss even for such a DC configuration.

Below are the thermal images of the experimental setup (Figure 5a) obtained in a steady state mode, as well as an image of the most preferable disinfection chamber (Figure 5b).

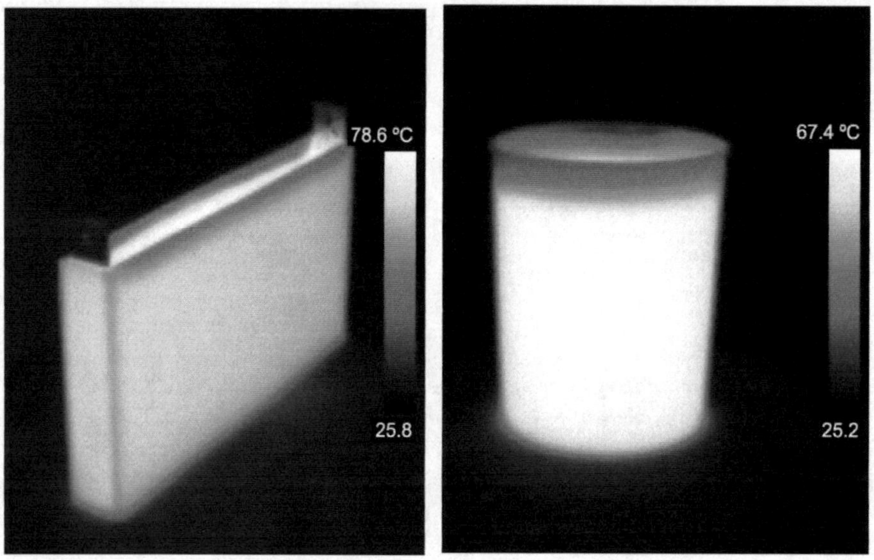

a) b)

Figure 5 – Thermal images: a) – rectangular experimental setup; b) – cylindrical container with the heated feed mixture

Figure 6 shows a thermal image with the distribution of the temperature field through the chamber wall.

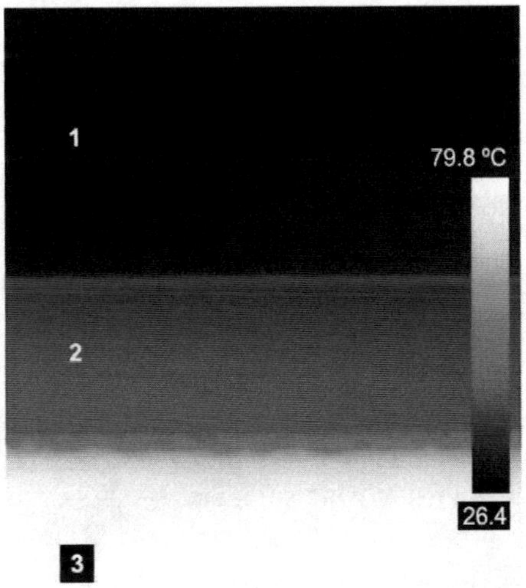

Figure 6 – Thermal image of temperature distribution through the partition wall: 1 – environment (air); 2 – dielectric chamber wall; 3 – heated feed mixture

The studies carried out have shown that with a decrease in the surface area of the enclosing walls of the DC unit, the heat loss due to heat transfer turns out to be negligible. At the same time, ensuring the lowest heat losses without the use of additional heat-insulating materials enclosing the unit is possible when using DC of cylindrical shapes.

Conclusion

1. The analysis of the mechanism of the appearance of the temperature gradient was carried out, as a result of which it was found that the temperature field is uniform throughout the entire volume of the feed mixture, an exponential decrease in temperature is observed only in the regions adjacent to the walls of the dielectric chamber, cooled from outside by air.

2.A diagram was provided forthe experimental setup with a description of the physical parameters of the heating chamber at a supply voltage of 150 V.

3. Experimentally, the relation between the temperatures and the heating time was established for points inside the volume of the feed mixture and on the outer wall in a transient mode of heat exchange, the values of the temperature head were determined.

4. Based on the equations of thermal conductivity, the heat transfer coefficients were calculated for the areas of contact between the media (feed mixture - environment), and the values of temperatures at these boundaries were determined.

5. The dependence of the growth of heat losses for a non-stationary mode of heat exchange in the process of heating the feed mixture is established, the values of which do not exceed 4% of the energy spent on heating the mixture.

6. The presented thermal images show the distribution of the heat flow for different configurations of dielectric chambers of the EC unit: cylindrical and rectangular parallelepiped.

References

[1] Gorshkov A V and Prosviryakov E Y 2020 Nonstationary laminar Bénard-Marangoni convection for Newton-Richmann heat exchange. In AIP Conference Proceedings (Vol. 2315, No. 1, p. 050010). AIP Publishing LLC.

[2] Dorozhkin M.V., Khalina, T. M., Khalin, M. V., Analysis of the skin effect impact in the technology of disinfection of fodder mixtures by electric contact heating. International Journal on Technical and Physical Problems of Engineering (IJTPE), Issue 38, pp. 170-174, October 2020.

[3] Anne Huss, Roger Cochrane, Cassie Jones, Griffiths G. Atungulu, Physical and Chemical Methods for the Reduction of Biological Hazards in Animal Feeds, Food and Feed Safety Systems and Analysis, 10.1016/B978-0-12-811835-1.00005-1, pp. 83-95, 2018.

[4] F.T. Jones, K.E. Richardson. Salmonella in commercially manufactured feeds. Poult. Sci., 83 (2004), pp. 384-391

[5] Montecucco A, Buckle JR, Knox A R. Solution to the 1-D unsteady heat conduction equation with internal Joule heat generation for thermoelectric devices. Appl Therm. Eng. 2012;35(1):177–84.

[6] Qin Y, Brockett A, Ma Y, Razali A, Zhao J, Harrison C, Pan W, Dai X, Loziak D (2010) Micro□manufacturing: research, technology outcomes, and development issues. Int J Adv Manuf Technol 47:821–837. doi:10.1007/s00170-009-2411-2

[7] S. Dilmac, A. Guner, F. Senkal, S. Kartal, Simple method for calculation of heat loss through floor/beam-wall intersections according to ISO 9164, Energy Conversion and Management, Vol. 48, Issue 3, (2007), pp. 826-835

[8] Mamun A. A., Chowdhury Z. R., Azim M. A., and Maleque M. A., Conjugate Heat Transfer for a Vertical Flat Plate with Heat Generation Effect, NAMC, vol. 13, no. 2, pp. 213-223, Apr. 2008.

[9] M. Miyamoto, J. Sumikawa, T. Akiyoshi, T. Nakamura, Effects of axial heat conduction in a vertical flat plate on free convection heat transfer, International Journal of Heat and Mass Transfer, 23, pp. 1545–1553, 1980

[10] Kundu, Balaram, and Kwan-Soo Lee., Exact analysis for minimum shape of porous fins under convection and radiation heat exchange with surrounding. International Journal of Heat and Mass Transfer 81 (2015): 439-448.

[11] De Lima, J. A. S., and J. Santos., Generalized Stefan-Boltzmann law. International Journal of Theoretical Physics 34.1 (1995): 127-134.

Development of a Range of Modified Multifunctional Residual Current Systems and Their Industrial Implementation

O K Nikolsky, T M Khalina
Russia, Barnaul, AltSTU

The developments of the AltSTU have been reviewed in the area of creating a new technology for preventing technogenic hazards based on the residual current devices. The residual current devices are intended for protecting people from electric shock in case of contact with conductive parts of the electric appliances and shall facilitate reduction of fire risks caused by a prolonged flow of leakage currents and fault currents resulting from them. The results of creating different modifications of protective trip circuits and their industrial use are provided.

The transition of Russia to power supply systems with separating working and protective neutral conductors (TN-C-S and TN-S) created the technical prerequisites for the massive use of residual current devices. The high electrical protective efficiency of the RCDs, due to the improved characteristics in terms of speed and sensitivity, compared to conventional means of electrical protection, imposes increased requirements for noise immunity to exclude false alarms and interruptions in the power supply to consumers, which led to the need for experimental studies of the parameters of power grids [1-3]. For this purpose, AltSTU created special measuring equipment, with the help of which the natural leakage currents of internal electrical networks and electrical receivers, short circuit currents, touch voltages were determined, the influence of external factors (rainfall, atmospheric and switching overvoltages, etc.) on the operation of RCDs was analyzed. ... In addition, studies were carried out on the influence of higher harmonics of the load current of protected lines, as well as currents induced by high-voltage power lines passing through rural areas [4, 5]. The results of the studies performed made the basis for the formation of a series of modified RCDs and the preparation of a design and technological base for their production in three main areas [6, 7].

1. Creation of multifunctional electronic protective devices.
2. Creation of protective devices for individual use of electronic design.
3. Creation of hybrid electrical protections with an electromechanical converter.

1. Multifunctional device for electrical and fire safety shutdown of increased power with overvoltage protection UZK-01 is intended for:

- protecting people and animals from electric shock;
- preventing fires and ignitions caused by leak currents, power surges, and short circuits;
- isolating electric receivers in case of overvoltages;
- measurement and alarm in case a setup value of residual current has been exceeded.

UZK-01 is a three-pole, four-wire autonomous residual current device controlled by a differential current, with discrete regulation of the magnitude of the tripping currents, with built-in protection against overcurrents and overvoltages, operating in alternating current electrical systems with a grounded neutral (Figure 1).

UZK-01 differs in the form of the differential current into types: AC - responsive to sinusoidal alternating current; A - responsive to sinusoidal alternating current and pulsating direct current.To ensure the possibility of serial installation, UZK-01 are distinguished by: UZK-01-S - selective; UZK-01 - non-selective.

Figure 1 – Appearance of RCD

Figure 2 – Arrangement of component parts of the RCD

UZK – 01 has the following advantages compared to its rivals:

1. Increased value of the rated current - 125 A with the prospect of its further increase to 250A (for existing analogues - 63A), which expands the scope and allows installing devices at the input of electrical installations of a production facility.

2. Multistage discretely adjustable setting of operation, which allows the device to be adapted to a real electrical network.

3. Presence of a pre-emergency mode alarm, which increases the reliability of the facility's power supply.

4. The ability to monitor the state of the insulation of the protected network, which increases the level of operation of electrical installations.

5. Overvoltageprotectioncapability.

Thus, UZK-01 is a universal residual current device with a wide range of applications, superior in functionality to well-known foreign analogues. During the tests at the Altai Geophysical Plant, the products were found to meet the specifications of ILKYu 425.328.001 TU in their parameters.

Testing schedule included measurement of the following parameters:

– the value of the rated breaking differential current I_n set by the setting switch;

– the value of the rated non-breaking differential current in accordance with the set position of the switch of the protection operation settings;

– operation of the differential current detection alarm in accordance with the set position of the protection operation settings switch;

– operation of the differential current detection alarm according to the set position of the setup tripping switchin the leakage current monitoring mode;

– protection operation time for all setups at operating voltage and presence ofthe residual current of $2I_n$;

– operation at overvoltage and overvoltage tripping operation.

Trial operation of UZK-01 products at residential and civil facilities has shown that residual current devices are effective and reliable means of electrical protection in low voltage electrical installations. To facilitate the connection of the product, the design of the unit should provide for the possibility of changing the location of the leakage current sensor - zero sequence differential current transformer (Figure 2).

2. Electronic Residual Current Device

A key component of the RCD is a leakage current sensor that is a differential current transformer (DCT). Its operation relies upon the induction of EMF in the toroidal winding, covering the conductor carrying the current measured. The current winding consists of two conductors for a single-phase network and four for a three-phase network, passing inside a toroidal core. In this case, the operating current can be dozens and hundreds of amperes, therefore the current conductor must have a sufficiently large cross-section, which does not allow the primary winding to be made of several turns. The measuring (secondary) winding is wound with an insulated wire so that the insulation between the first and last turns, as well as between the terminals, must withstand voltages up to 1000 V.

In order to implement the principle of constructing noise-immune protective shutdown circuits, which, at significant rated load currents (up to 125 A), react to very small current leakages not exceeding 6 or 30 mA, it is necessary to develop a sensor (DCT) with the following properties:

1. The material for the magnetic circuit must have low values of coercive force and high magnetic permeability, as well as have a minimum residual induction to ensure the stability of the sensor parameters (DCT) after exposure to overcurrents. Magnetic materials for a DCT magnetic circuit operating in an alternating magnetic field must have a narrow hysteresis cycle and a large slope of the V (H) characteristic in weak fields.

2. The design of the DCT should provide noise immunity under the influence of EMF unbalance and other destabilizing factors, the protection operation setting should not go beyond the specified limits.

3. The inductance of the measuring winding together with the input capacitance of the device must provide resonance at the frequency of the supply network and have sufficient resistance to limit the current that occurs when the device is disconnected in the event of a short circuit in the network.

The developed and manufactured DCT sample with the GAMMAMET alloy core meets all the above requirements and provides an output voltage of 0.008; 0.026 and 0.25 V at leakage currents of 0.006, 0.01, 0.03 and 0.3 A, respectively. The transformer allows operation and retains its parameters after exposure to overcurrents up to 1500 A, while the error does not exceed two percent.

The reasons for the emergence of EMF imbalance in a differential current transformer are currently quite fully investigated. The main ones are the imperfection of the geometric design of the DCT and the asymmetric arrangement of current-carrying conductors in relation to the magnetic circuit. The EMF imbalance value depends on the load current, the ratio of the external and internal radii and the magnetic permeability of the toroidal core, the degree of displacement of the current-carrying conductors relative to the geometric center of the magnetic circuit. The most effective methods of EMF imbalance reduction are use of materials with high magnetic permeability for the magnetic circuit; the use of toroidal cores with the greatest possible multiplicity of dimensions of the outer and inner diameters; reduction of the cross-section of the secondary winding conductors; shielding the core with screens made of soft magnetic materials. Meeting these requirements allows at the same time to significantly reduce the influence of external magnetic fields. The filters of the higher harmonics of the response delay system in the DCT signal converter allow to avoid false RCD trips as a result of the influence of higher harmonics of the load current, as well as currents induced by high-voltage lines and lightning discharges.

The above methods of EMF imbalance reduction and increase noise immunity against false operation can be used in the RCDs with electronic amplifiers. The DCT here plays an extra role of a voltage transformer connected to the electronic amplifier with a high input resistance. The design of such DCT should provide for a large number of secondary winding turns to get a required useful signal. However, one should bear in mind that the increase in the number of turns results in the increase in the RCD sensitivity to the interference.

The studies have shown that the differential current transformers meet the following requirements:

– magnetic core material has minimum residual induction;

– the minimum value of the magnetic permeability corresponds to the intensity of the magnetic field created by a current of not more than 0.01 A;

– the inductance of the secondary winding of the transformer lies within 0.5 - 4 Hz, and the capacitance of the capacitor connected at the input of the device is 2 - 20 mcF, thereby limiting the current in the winding to 2.5 A and a resonant frequency of 50 Hz;

– the primary winding wire can withstand a current of up to 1500 A, has a mechanical fastening that prevents the insulation from moving, and has a dielectric strength of up to 1500 V;

– the wire of the secondary winding of the transformer is capable of passing a current of up to 2.5 A for a long time, and the insulation between the terminals of the winding can withstand voltage up to 500 V;

– the design of the transformer provides an additional winding in the secondary circuit (≈10 turns) to carry out the function of monitoring the insulation of the electrical network.

The use of such a DCT design made it possible to create electronic RCDs operating at rated load currents up to 125 A and a rated breaking differential current up to 30 mA.

3.A Hybrid Electromechanical RCD

One of the promising directions in the development of residual current devices in Russia is the creation of electromechanical devices that have a number of advantages over electronic ones. However, domestic production of RCDs is limited to electronic products and semi-finished products assembled from foreign components (Austria, China). The reason for this is the lack of technology for creating high-precision polarized relays with a kinematic amplifier to control the RCD switching unit.

The solution to the issue of creating electromechanical RCDs is possible with the participation of an enterprise that manufactures circuit breakers and makes their revision in order to be used as an RCD switching unit. In this case, the implementation of various options is possible determined not only based on the optimal layout, but also on the parameters of the electromechanical converter, as well as the sensor - differential current transformer. Depending on the option of choice, various methods of compensation for the unbalance current should be used as well as stabilization of characteristics caused by the possibility of magnetizing the magnetic circuit of the DCT by short-circuit currents, detuning from the influence of the three-fold external magnetic interference and higher harmonics leading to false alarms, etc.

Thus, the key tasks of the investigation in creation of a new RCD type are:

–development of an electromechanical converter and differential current transformer;

– development of a manufacturing procedure for a hybrid RCD comprising a magnetic and electric relay (instead of an electronic and kinematic amplifier) and a triac controlling a shunt release of an automatic circuit breakerconnected with the sensing unit of the RCD.

These tasks have been solved during preparation of RCD manufacture at the Divnogorsk plant of low-voltage equipment (Krasnoyarsk Territory).

Study of the Electromechanical Converter. Abroad [8], the most common design options for electromechanical converters are bipolar and unipolar circuits (Figure 3). In both systems, the movable anchor is held in its original state by a permanent magnet, which, among other things, overcomes the resistance of the opposing spring. The current flowing through the yoke changes or weakens the constant magnetic field. In this case, the anchor is detached from the magnetic circuit under the action of the spring and, by means of a pusher, acts on the opening mechanism of the circuit breaker.

IOP Conf. Series: Materials Science and Engineering 1211 (2022) 012020 doi:10.1088/1757-899X/1211/1/012020

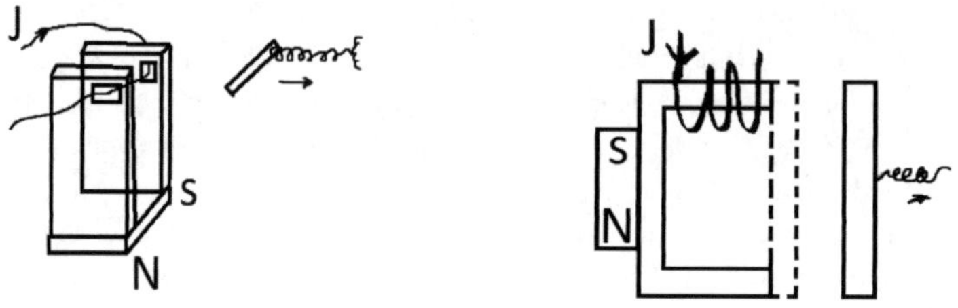

Bipolar circuit Unipolar system

Figure 3 – Options of electromechanical transducer with holding magnet

The difference between the options lies in the fact that in the bipolar circuit, the off state is provided for any direction of the current, and in the unipolar circuit, only for one direction, corresponding to the weakening of the permanent magnet field.

In the proposed relay design, a bipolar circuit is used, shown in Figure 4. The action of such a relay is based on the interaction of the magnetic field of permanent magnet 5 and an alternating current (50 Hz) in winding 4, located on anchor 3. Rocker arm 6 and the armature act as a magnetic flux conductor, created by a permanent magnet, which ensures the attraction of the rocker arm to the anchor. By turning the permanent magnet around its axis, it is possible to change the tangential component of the magnetic flux flowing through the magnetic circuit, and, thereby, to regulate the force of attraction of the rocker arm to the armature, setting the required relay sensitivity, which determines the response setting. When a leakage signal arrives at the relay coil from the output of the differential current transformer, the armature is demagnetized, the rocker arm, under the action of spring 7, leans back and through pusher 2 closes the normally open contacts of the relay. The function of the kinematic amplifier is performed by a triac that controls the shunt release of the circuit breaker and is included in a separate unit of the hybrid residual current device.

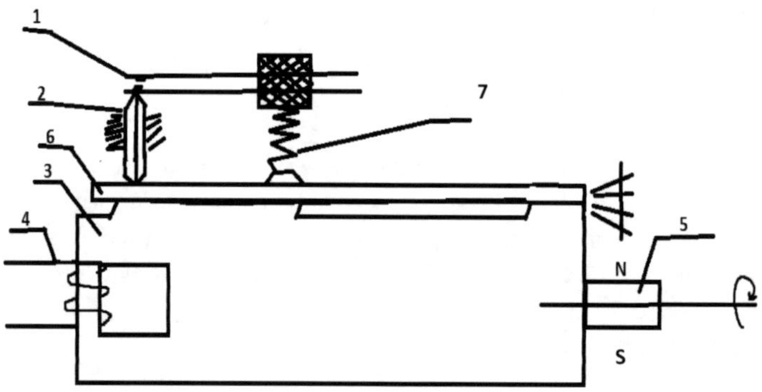

Figure 4 – Magneto electric relay design

Considering that the main structural elements of a magnetic relay are a permanent magnet, a magnetic circuit and a spring, which determine the sensitivity and reliability of the RCD, we carried out research on the selection of raw materials and the development of technology for the manufacture of parts and their assembly. The main requirements for the material of the magnetic circuit are high magnetic permeability and a narrow hysteresis loop, taking into account operation in weak magnetic fields (0.5-2 A/m). The studies carried out have shown the possibility of using permalloy type 79NM (GOST 101160-75) for the magnetic circuit.

The spring of the magnetic relay is the most important structural element and must meet the requirements to ensure the stability of its dynamometric characteristics over time. Failure to comply with these requirements may lead to a change in the RCD trip setting during operation, which is unacceptable. Therefore, the use of silicon-vanadium steels with a high modulus of elasticity has been proposed as a spring material. The tests also showed the possibility of using steels with a lower elasticity, in particular, a wire spring made of 65G steel (GOST 9384-80), provided that the wire from which the spring is wound is used horizontally and in the direction opposite to the winding.

Study of the Differential Current Transformer. In residual current devices with an electromechanical converter, the DCT is used as a current transformer connected to a load with a low input resistance. Since the number of turns of the secondary winding of the DCT in this case is very small (from several turns - up to several dozens), the sensitivity to the influence of interference sources is significantly reduced. Therefore, the practical implementation of measures to prevent false alarms of such RCDs is greatly simplified. One of the issues of stabilizing the characteristics of the RCD is the elimination of residual magnetization of the DCT core by single-phase short-circuit currents, leading to a change in the sensitivity and unstable operation of the protection. The reason for the change in sensitivity is the change in the slope of the hysteresis loop of the DCT core after magnetization of the short-circuit current [9].

An effective way to stabilize the characteristics is by connecting a pre-charged capacitor in parallel with the secondary winding of the DCT after each operation of the RCD. At the same time, an auxiliary oscillatory process occurs in the LC-circuit formed by a capacitor with a DCT winding, which ensures demagnetization of the DCT along an anhysteresis-free curve, that is, the establishment of its operating point in the initial section of the magnetization curve. The nominal value of the capacitance of the capacitor is chosen so that the resonant frequency of the LC-circuit is higher than the industrial frequency of the protected network. This is necessary to eliminate the influence of resonance at the power frequency during the operation of the residual current device.

Taking into account the results of the research carried out, a residual current device has been developed, the structural, schematic diagrams and general view of which are shown in Figures 5, 6 and 7.

The device contains differential current transformer 1, to the output of which electromechanical converter 2 is connectedlinked toswitching unit 3 containing an electromechanical contact 4 connected in series with current-limiting resistors 5 and 6 and having the ability to control semiconductor switch 7 based on a triac; switching unit 3 connected to actuator 8. The actuator is a circuit breaker with a shunt release, which trips when voltage is applied to the release coil from an external power source.

IOP Conf. Series: Materials Science and Engineering 1211 (2022) 012020 doi:10.1088/1757-899X/1211/1/012020

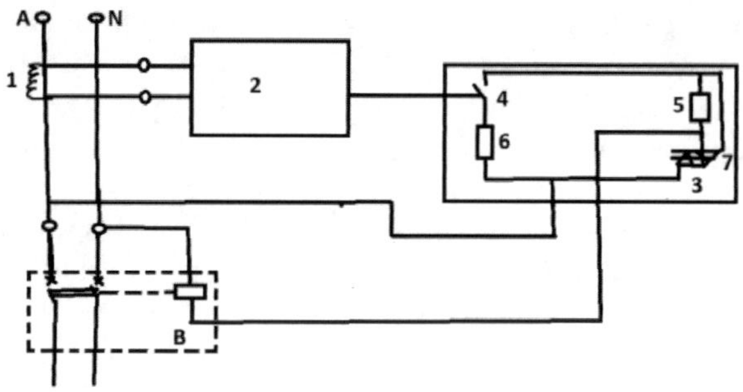

Figure 5 – Structural diagram of RCD-ЭЛНИС

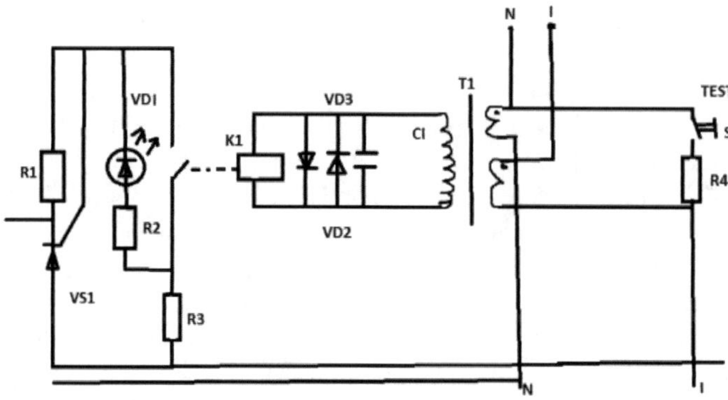

Figure 6 – Electric wiring diagram of RCD-ЭЛНИС

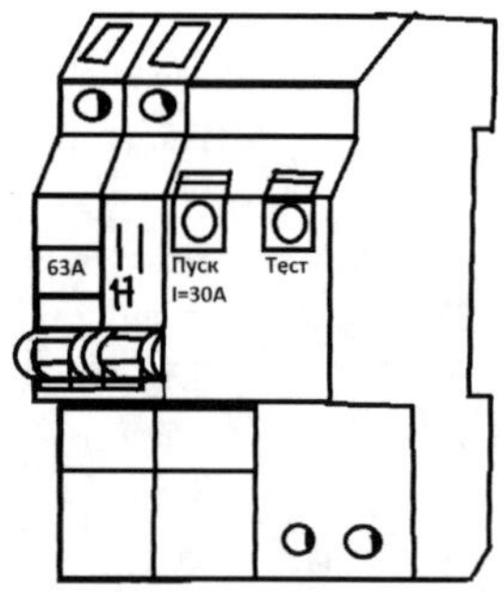

Figure 7 – General view of residual current device RCD-ЭЛНИС (mockupspecimen)

Industrial Use of a Hybrid RCD with an Electromechanical Converter. The successful test results of the developed RCD made it possible to launch the production of the first hybrid RCDs, combining an electronic amplifier based on a semiconductor switch with an electromechanical converter from domestic components. The RCD is produced on the basis of the VA 61-29 circuit breaker (Figure 7).

The product is a single-phase two-wire stationary autonomous residual current device with built-in overcurrent protection controlled by differential current with a fixed magnitude of the operating current, operating in alternating current power grids with a frequency of 50 Hz with a rated voltage of 220 V and a load current of up to 63 A. The RCD is characterized by increased reliability. The calculated MTBF of the product is more than 30 years, which is 3-4 times higher than that of an electronic RCD. The volume of production since the beginning of production has amounted to more than 50,000 units [10].

References

[1] Abdullah M. Z., Ariffin R. Power quality analysis of Residual Current Device [RCD] nuisance tripping at commercial buildings //2013 IEEE Symposium on Industrial Electronics & Applications. – IEEE, 2013. – C. 122-125.

[2] Czapp S. Protection against electric shock using residual current devices in circuits with electronic equipment //Elektronika ir Elektrotechnika. – 2007. – T. 76. – №. 4. – C. 51-54.

[3] Shopov Y., Filipova-Petrakieva S., Boychev B. Investigation of residual current devices in high frequencies //2018 10th Electrical Engineering Faculty Conference (BulEF). – IEEE, 2018. – C. 1-3.

[4] R.N. Karyakin, Standards for the construction of safe electrical installations. - M.: Energoservis, 2000. - 453 p.

[5] Parise G., Annibaldi S., Martirano L. Toward alignment of safety levels concerning electrical installations in the EU // IEEE Industry Applications Magazine. – 2001. – T. 7. – №. 3. – C. 73-82.

[6] Biegelmeier G. Kannder Fehlerstromschutzschalter die Technik des Beruhrungsspannungsschutzes revolutionieren? “Electrotechnik und Maschinendan”, 1954. №4.

[7] Rotter K. Elektrosehuts-Felten Guibleaume. Fabrik Elektriseher Apparate Aktiengesellschaft, 1990.

[8] TEC Standart 364-4-41 / Electrial installations of buildings. Part 4: Protection for safety. Chapter 41: Proctionagainst electric shoks.1992 – 10.

[9] Stanbury M., Djekic Z. The impact of current-transformer saturation on transformer differential protection // IEEE Transactions on Power Delivery. – 2014. – T. 30. – №. 3. – C. 1278-1287.

[10] A. I. Logvinov, A.P. Rodin, Experience of batch production of residual current devices in Russia // Newsletter of Polzunov AltSTU- Barnaul, 2000. No.3.

Electroresistive method for monitoring the technical condition, quality, reliability and safety the supports of rolling of electrical machines

K V Podmasreryev[1], **V V Markov**[1], **V V Mishin**[1], **A V Selikhov**[1] **and N V Uglova**[1]
[1] Oryol State University named after I.S. Turgenev,
95 Komsomolskaya Street, Oryol, Russia

Abstract. The necessity of monitoring the technical condition of the rolling supports of electric machines has been substantiated. It is proposed to use the electrical resistance of the bearing as an indicator of the technical condition the rolling support. The results of mathematical modeling of electrical resistance in the form of a function of resistance from factors of the internal environment of the bearing and modes of its assembly and operation in a rolling bearing are presented. An electroresistive method for monitoring the technical state the rolling support is proposed, which differs from the known methods by original algorithms for collecting information about the state of bearing parts, experimental studies have been carried out to confirm the efficiency of this method.

1. Introduction

The quality, reliability and safety of electrical machines are determined by their technical condition, which depends on the degree to which the dimensions and parameters of the elements of the electrical machine correspond to their nominal values. Rolling supports play an important role in the formation of the technical condition of an electric machine, since they experience the most intense mechanical loads. The state of the rolling support is determined by the parameters of its main element - the rolling bearing. Thus, to determine the technical condition of the rolling bearing of an electric machine, it is necessary to assess the quality of the bearings that make up this bearing.

Rolling bearings are standard products and are of high quality, reliability and durability as delivered; many modern bearings have an integrated self-lubrication system with artificial intelligence. However, the high quality of the bearing does not provide complete confidence in the performance of the rolling bearing, in which the given bearing is assembled, since its actual state in the bearing is determined by the assembly and operating modes of the bearing. Factors in the assembly and operation of the rolling element can significantly change the condition of the bearing and the electrical machine as a whole.

Among the known methods for monitoring rolling bearings, electrical methods are distinguished. These methods are universal, relatively simple to implement, and inertialess. A variety of electrical methods for monitoring rolling bearings are electroresistive methods. They are based on the physical phenomenon of the dependence of the electrical resistance of the bearing on the factors of its internal environment, assembly modes and operation of the rolling support. Electroresistive methods for monitoring rolling bearings became known in the second half of the 20th century, and information about them was published in many review works, in particular [1, 2].

A well-known group of electroresistive methods for monitoring rolling bearings are methods for monitoring the time of contact of the metal surfaces of rings and rolling elements of bearings. These methods were first developed by scientists S.F. Korndorf and K.V. Podmasreryev in the period from 1979 to 1986, were subsequently developed in the works of their scientific school, and the basic diagnostic parameter received the abbreviation "NIV" - "normalized integral time" [1].

The use of the NIV parameter is limited by two conditions: the mode of mixed lubrication in the bearing and the time of determining the average value of the parameter [1, 3]. A more universal indication of the condition of a rolling bearing is electrical resistance.

The purpose of this work is to describe the results of the development of a method for monitoring the quality of the support by the parameters of its electrical resistance. The parameters of electrical resistance make it possible to carry out a comprehensive assessment of the state of the rolling support, as well as to identify individual defects of its elements in arbitrary lubrication modes.

2. Theoretical basis for electroresistive monitoring of rolling supports

The theoretical basis for the electroresistive method of rolling supports monitoring is a mathematical model of the electrical resistance of the bearing. A rolling bearing is a complex technical system in which mechanical, hydrodynamic and electrical factors are combined. Modelling the electrical resistance of a bearing is based on the use of several fundamental theories, a detailed description of which is presented in [3]. Mechanical factors are modelled in accordance with the theory of contact of real surfaces of the scientific school of N.B. Demkin [4]. The state of the lubricating film is described in accordance with the hydrodynamic theory of lubrication of the scientific school of D.S. Kodnir [5]. The electrical resistance of the friction contact is modelled in accordance with the theory of electrical conductivity of the contact between two rough bodies of the scientific school of A.V. Bely and N.K. Myshkin. The connections between individual frictional contacts within the contact spots of the bodies and bearing rings are modelled by the Nyack distribution [6, 7].

The theoretical basis for the electroresistive method of rolling bearing monitoring is a mathematical model of the electrical resistance of the bearing. A rolling bearing is a complex technical system in which mechanical, hydrodynamic and electrical factors are combined. The basic structural element of a rolling bearing, as a tribological object, is an elementary frictional contact, and the transition from the resistance of a frictional contact to the resistance of the bearing as a whole is carried out in accordance with the equations of M.P. Kovalev and N.Z. Narodetsky [8]. For frictional contact, the retraction resistance R_{ST} and the lubricating film resistance R_{SP} are taken into account, determined through the resistivity of the contacting surfaces (ρ), tunnel surface lubricating films (ρ_{P}), boundary film (ρ_{GP}) and bulk lubricant (ρ_{SM}). The calculation takes into account the thickness of the lubricating film in contact (h) within the area (S_h), the radii of the contact spots for the inner (V) and outer (N) bearing rings ($r_{\mathrm{V(N)}}$) and the number of contact spots within the friction zones (n_{P}):

$$\begin{cases} R_{ST} = \dfrac{\rho}{2 \cdot n_P \cdot r_{\mathrm{V(N)}}}, \ h<3\ nm; \ R_{SP} = \dfrac{\rho_P}{\pi \cdot n_P \cdot r_{V(N)}^2}, 3 \le h < 10 nm; \\ R_{SP} = \rho_{GP} \cdot \dfrac{h}{S_h}, 0{,}01 \le h < 1\,\mu m; \ R_{SP} = \rho_{SM} \cdot \dfrac{h}{S_h}, 1 \le h \le 30\,\mu m. \end{cases} \quad (1)$$

The system of equations (1) makes it possible to take into account the possible modes of lubrication in the bearing: dry, boundary and liquid friction. It is the basis of the mathematical model.

The mathematical model of the bearing electrical resistance is represented by the system of equations (1-10). Equations (2) and (3) serve to calculate the contact resistance:

$$R_{\mathrm{ST}} = a_R \cdot \Psi_R(h); \ R_{\mathrm{SP}} = R_{\mathrm{ST}} \cdot \frac{\rho_{\mathrm{SM}}}{\rho_{\mathrm{M}}} \cdot \frac{h_0}{r_{\kappa}} \cdot \frac{A_r}{A_c - A_r}; \ a_R = \frac{\rho_{\mathrm{M}}}{2 \cdot N \cdot \sqrt{\beta \cdot d}}; \quad (2)$$

$$\Psi_R(h)=\left|L\left(\frac{d}{R_q}\right)\cdot\left[1-L\left(\frac{d}{R_q}\right)\right]+\frac{R_q}{d\sqrt{2\pi}}\cdot\left[\frac{1}{2}-L\left(\frac{d}{R_q}\right)\right]\cdot exp\left(-\frac{d^2}{2R_q^2}\right)-\frac{1}{4}\right|^{-0,5};\ \beta=\sqrt{r_{\text{pr}}\cdot r_{\text{pop}}}\ . \quad (3)$$

To calculate the retraction resistance R_{ST}, the following must be specified: auxiliary variables of the electrical resistance function a_R and $\Psi_R(h)$; electrical resistivity of lubricant (ρ_{SM}) and metal surfaces (ρ_M); the smallest thickness of the lubricating film within the contact patch (h_0); actual (A_r) and contour (A_c) contact areas. The following parameters are used in the formulas for the functions a_R and $\Psi_R(h)$: integral Laplace function $L(d/R_q)$; radius of curvature of microroughnesses (β), taking into account the radii of irregularities in the longitudinal (r_{pr}) and transverse (r_{pop}) directions; nominal clearance between the rolling element and the bearing ring (d); the number of irregularities (N) of the profile within the nominal contact area (A_a) of the rolling elements and bearing rings; root-mean-square deviation of the height of irregularities from the midline of the profile within the base length (R_q). The main factor that forms the electrical resistance of a contact is the actual contact area (A_r), defined as a certain part of the contour contact area (A_c). Functions for calculating A_r and A_c are given by equations (4) and (5):

$$A_r=\pi\cdot N\cdot\beta\cdot d\cdot\left[L\left(\frac{d}{R_q}\right)+\frac{R_q}{d\cdot\sqrt{2\pi}}\cdot\exp\left(-\frac{d^2}{2\cdot R_q^2}\right)-\frac{1}{2}\right];\ d=R_p-\delta;\ N=\frac{A_c}{S_{m1}\cdot S_{m2}}; \quad (4)$$

$$A_c=\pi\cdot n_a\cdot n_b\cdot\sqrt[3]{\left(\frac{3\cdot F_r}{E\cdot\Sigma\rho}\right)^2}-A_{\text{D}};\ \delta=h-d=\sqrt[3]{\Sigma\rho}\cdot\left[\frac{3}{2}\cdot\frac{2\cdot K}{\pi\cdot n_a}\cdot\sqrt[3]{\frac{1}{3}\cdot\left(\frac{1-\eta^2}{E}\right)^2\cdot F_r^2}\right]. \quad (5)$$

To calculate A_r and A_c, the above parameters must be specified, as well as: the rolling element and ring approaching (δ); the height of the surface irregularities (h) and the dimensionless height of the irregularities (h_1); the area of the local defect on the surface (A_D); the step of the irregularities of the profile in the longitudinal (S_{m1}) and transverse (S_{m2}) sections; the reduced height of smoothing the irregularities of the profile of surfaces 1 and 2 ($R_p = R_{p1} + R_{p2}$); coefficients n_a, n_b and $(2\cdot K/\pi\cdot n_a)$, taking into account the shape of the contacting surfaces; geometric parameter of the sum of principal curvatures of contacting surfaces ($\Sigma\rho$); modulus of elasticity (E); Poisson's ratio (η); force of normal loading of frictional contact (F_r).

The equation of the contour contact area A_c allows one to take into account the effect on the electrical resistance of the contact of two types of surface shape deviations: local defects (holes, shells, risks, etc.) and macrodeviations (eccentricity, ovality, bearing ring cut). The influence of local defects is taken into account by introducing the parameter of the area of the local defect A_D into equation (5). The influence of macrodeviations is estimated indirectly through the sum of the principal curvatures of surfaces ($\Sigma\rho$):

$$\Sigma\rho=\Sigma\rho_0+\sum_{i=1}^{p}Q_k\cdot\sin\left(k\cdot\varphi+\varphi_k\right), \quad (6)$$

where $\Sigma\rho_0$ – is the sum of the main curvatures for the contact of surfaces with the nominal profile;

Q_k, φ_k – amplitude and phase angle of the k-th harmonic of macrodeviations;

k – harmonic number (1 – eccentricity, 3 – ovality, 3 – three-vertex cut);

p – is the limiting number of the considered harmonic.

The random function of the actual contact area A_r depends on the distribution law of the random variable of the dimensionless height of the irregularities of the contacting surfaces h_1, which is described by the Nyack distribution and is given by equations (7) and (8):

$$f(h_1)=\frac{\sqrt{3}}{2\pi}\cdot\left\{\begin{array}{l} h_1\cdot\sqrt{\frac{3(2\alpha-3)}{\alpha^2}}\cdot\exp\left(-C_1\cdot h_1^2\right)+\frac{3\sqrt{2\pi}}{2\alpha}\cdot\left(h_1^2-1\right)\cdot\left[1+\text{erf}\left(h_1\cdot\sqrt{\frac{3}{2(2\alpha-3)}}\right)\right]\times \\ \times\exp\left(-\frac{h_1^2}{2}\right)+\sqrt{\frac{2\pi\alpha}{3(\alpha-1)}}\cdot\left[1+\text{erf}\left(h_1\cdot\sqrt{\frac{\alpha}{2(\alpha-1)(2\alpha-3)}}\right)\right]\cdot\exp\left(-\frac{\alpha h_1^2}{2(\alpha-1)}\right)\end{array}\right\}. \quad (7)$$

$$h_1=\frac{h}{\sqrt{m_0}};\ \alpha=\frac{m_0\cdot m_4}{m_2^2};\ C_1=\frac{\alpha}{2\cdot\alpha-3}, \quad (8)$$

where m_0, m_2, m_4 – are the moments of the spectral density of the height of the irregularities.

Formulas (1) - (8) model the electrical resistance of the contact of one rolling body with the bearing rings. To calculate the resistance of the bearing as a whole, it is represented as a parallel connection of the contacts of all rolling bodies, and the contact resistance of each rolling body with the rings is determined by solving the problem of the distribution of the normal loading force between the rolling bodies of the bearing [2].

The compiled mathematical model of the electrical resistance of the bearing makes it possible to assess the effect on the electrical resistance of the rolling support: the roughness of the surfaces of the rolling elements and bearing rings, local defects and macrodeviations of the working surfaces of the bearing parts, operating modes of the rolling support.

3. Contents of the electroresistive method of rolling support monitoring

The essence of the electroresistive monitoring method is to assess the technical condition of the rolling support by the parameters of electrical resistance. The control takes place in two stages: averaged complex assessment of the technical condition; assessment of the shape and size of deviations in the shape of bearing rings loaded with local or circulating loading. The following are used as diagnostic parameters:

- parameter R_s – the arithmetic mean of the bearing resistance, averaged over a time equal to the period of rotation of its rolling ring;
- parameter K_n – is the ratio of the power of the harmonic component of the bearing resistance function versus time $R(t)$ at the information frequency to the power of the carrier harmonic of the function $R(t)$:

$$K_n=(\sigma_{RFi})^2/(\sigma_{R0})^2\ ;F_i=F_0\cdot M\ ;\ F_i=F_s\cdot z+F_0\cdot M\cdot L;\ F_i=F_s\cdot z+3\cdot F_0\cdot M\ , \quad (9)$$

where $(\sigma_{RFi})^2$ – power of the component function $R(t)$ at the information frequency F_i, W;

$(\sigma_{Rf0})^2$ – power of the component function $R(t)$ at the carrier frequency F_0, W;

F_0 – carrier frequency (frequency of rotation of the ring), Hz;

F_s – bearing cage rotation frequency, Hz;

M – macrodeviations shape factor (1 – eccentricity; 2 – ovality; 3, 4 ... n – n-order cut);

L – empirical coefficient of proportionality equal to 1 or 2.

The developed electroresistive method makes it possible to monitor the rolling support by several techniques, differing in the types of detected defects [2].

3.1 *Technique for assessing the general condition of a bearing in a rolling support.* This technique is based on the use of the R_s diagnostic parameter and includes the following actions:

- select the object of monitoring: select the electric motor, the rolling bearings of which are subject to monitoring, determine the type of rolling bearings in the bearings;
- the modes of operation of bearings in rolling bearings are assessed: normal load, rotation speed of the rings, type, brand and amount of lubricant;
- turn on the object and run in the bearings for 2 hours;
- measure the diagnostic parameter R_s for a time equal to 10 ... 50 periods of rotation of the bearing ring;

- based on the value of the R_s parameter, a conclusion is made about the lubrication mode in the bearing and the state of the rolling support (based on the results of comparing the measured value of the R_s parameter with the value acceptable for a given type of bearing and a given lubrication mode).

This technique is easy to implement, is carried out using relatively simple technical means [11] and is efficient in all lubrication conditions.

3.2 *Technique for assessing macrodeviations of rolling bearing parts with local loading*. This technique allows the quality of the rolling support to be assessed by the condition of the bearing ring under local loading. It is based on the use of the Kn diagnostic parameter and includes the following actions:

- select the object of monitoring: select the electric motor, the rolling supports of which are subject to monitoring, determine the type of rolling bearings in the supports;
- the modes of operation of bearings in rolling supports are assessed: normal load, rotation speed of the rings, type, brand and amount of lubricant;
- turn on the object and run in the bearings for 2 hours;
- connect the rolling support to a monitoring device that allows you to measure the R_s parameter for a time equal to 0.05 ... 0.1 of the rotation period of the bearing ring;
- the test bearing ring is divided into n sections;
- receive n readings of the diagnostic parameter R_s;
- build a function of dependence of the diagnostic parameter R_s on the angular coordinate of the normal load vector Fr and carry out its harmonic analysis;
- by the extrema of the harmonic components of the amplitude spectrum, the R_s functions determine the type of the main macrodeviations present on the bearing ring.

The advantage of this technique is the ability to control the durability of the rolling support based on the results of detecting macrodeviations of a locally loaded bearing ring and choosing a rational position of this ring relative to the load.

3.3 *Technique for evaluating macrodeviations of rolling support parts with circulating loading*. This technique is based on the use of the diagnostic parameter K_n and includes the following actions:

- select the object of monitoring: select the electric motor, the rolling supports of which are subject to monitoring, determine the type of rolling bearings in the support;
- the modes of operation of bearings in rolling support are assessed: normal load, rotation speed of the rings, type, brand and amount of lubricant;
- turn on the object and run in the bearings for 2 hours;
- connect the rolling support to a monitoring device that allows you to measure the parameter K_n at information frequencies calculated by the block of formulas (9);
- based on the results of comparing the value of the parameter K_n with reference data for a given type of bearing, the type, value and orientation of macrodeviations are determined.

The advantage of this technique is the ability to control the durability of the rolling support based on the results of detecting macrodeviations of a circulatingly loaded bearing ring and choosing a rational frequency of rotation of the electric motor shaft.

4. Analysis of experimental data

In order to experimentally test the performance of the proposed electroresistive monitoring method and the declared capabilities of the methods for assessing the state of the rolling support, experimental studies were carried out. Unlike the well-known research results [1, 2], special attention was paid to the operability of the method in the liquid lubrication mode, therefore the bearing was lubricated not with transformer oil, which forms a mixed lubrication mode, but with lubricants that provide a strong and continuous lubricating film, in which the parameter NIV is inoperative: with TAD-17 transmission oil and plastic lubricant "Litol-24".

The object of research was a rolling support with a 1000900 type bearing. An experimental rolling support was installed in a well-known setup, which allows simulating macrodeviations of bearing rings in the range of values from 0 to 50 μm, setting the shaft rotation frequency in the range of 0 to

3000 min^{-1}, normal load in the range of values from 0 to 500 N with local and circulating loading of the bearing rings. A well-known automated system for collecting and analyzing data during tribomonitoring SADT-1 was used as a means of measuring the resistance of the rolling support.

In order to increase the reliability of the experimental data, the research object was thoroughly prepared for the experiment – the test bearing was washed with kerosene, dried for 20 hours, and run-in for 2 hours. During the experiments, stable values of the rolling support operating modes were set: the normal load value was 100 N, the volume of lubricant varied in the range from 8 to 40 mm^3, the shaft rotation frequency was 6000 min^{-1} when examining a locally loaded ring and 4500 min^{-1} for circulatingly loaded ring.

Experimental data have confirmed the efficiency of the electroresistive monitoring method and the declared capabilities of the methods for assessing the state of the rolling support.

Figure 1 shows the results of assessing the general condition of the bearing in the rolling support using the diagnostic parameter R_s.

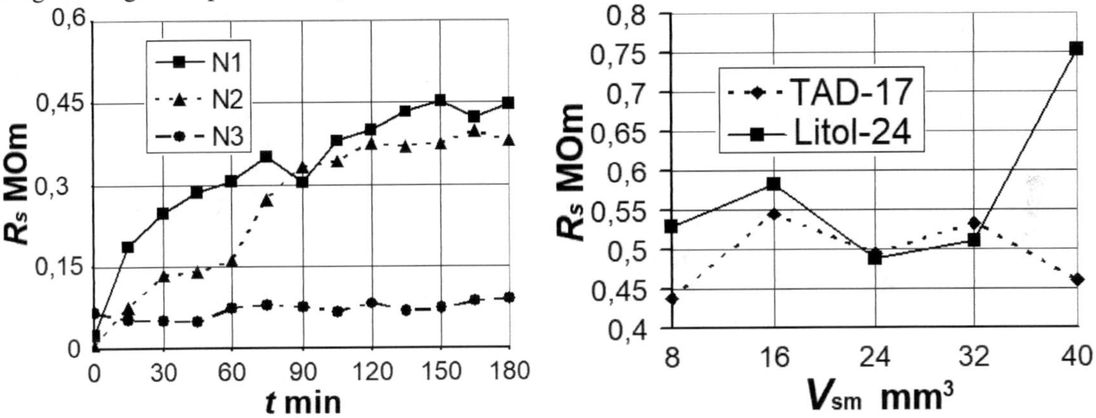

a) b)

a – operating mode influence; b – influence of the type and amount of lubricant

Figure 1. Influence of the operating mode and lubricant on the parameter R_s

Figure 1, a show the dependence of the parameter R_s on time during the running-in process of the bearing. Dependencies N1 and N2 correspond to new bearings, and dependence N3 – to a worn bearing in the limiting state. Figure 1, b shows the dependence of the R_s parameter on the type of lubricant and its amount. It can be seen that the values of the parameter R_s significantly depend on the amount of lubricant in the bearing, and the dependences are individual for each type of lubricant.

Figure 2 shows the results of evaluating the macrodeviations of rolling bearing parts with local loading according to the diagnostic parameter R_s.

Figure 2, a show the dependence of the R_s parameter on the ovality of the locally loaded outer ring of the bearing lubricated with TAD-17 semi-fluid transmission oil. It can be seen from the graph that with an increase in the ovality value from 5 to 40 μm, the value of the average electrical resistance of the bearing decreases from 0.65 to 0.45 MOhm. A decrease in the electrical resistance of the bearing with an increase in ovality indicates the sensitivity of the R_s parameter to macrodeviations of the bearing rings, and the relatively small dynamic range of the R_s parameter indicates the good lubricating properties of a semi-liquid gear oil.

Figure 2, b shows the dependence of the R_s parameter on the ovality of the locally loaded outer ring of the bearing lubricated with Litol-24 grease. It can be seen from the graph that with an increase in ovality from 5 to 40 μm, the value of the average bearing resistance decreases from 0.7 to 0.2 MOhm. At the same time, the nature of the dependence confirms the sensitivity of the R_s parameter to macrodeviations of the bearing rings, and a wider dynamic range of the R_s parameter indicates a lower efficiency of the grease compared to semi-liquid gear oil. Particularly noteworthy is the section of the graph in Figure 2, b with the ovality of the outer ring Q = 30 μm. In this area, there is a significant scatter in the values of the parameter R_s, which indicates a sharp deterioration in the lubrication regime and the degradation of the lubricating properties of the material at high loads.

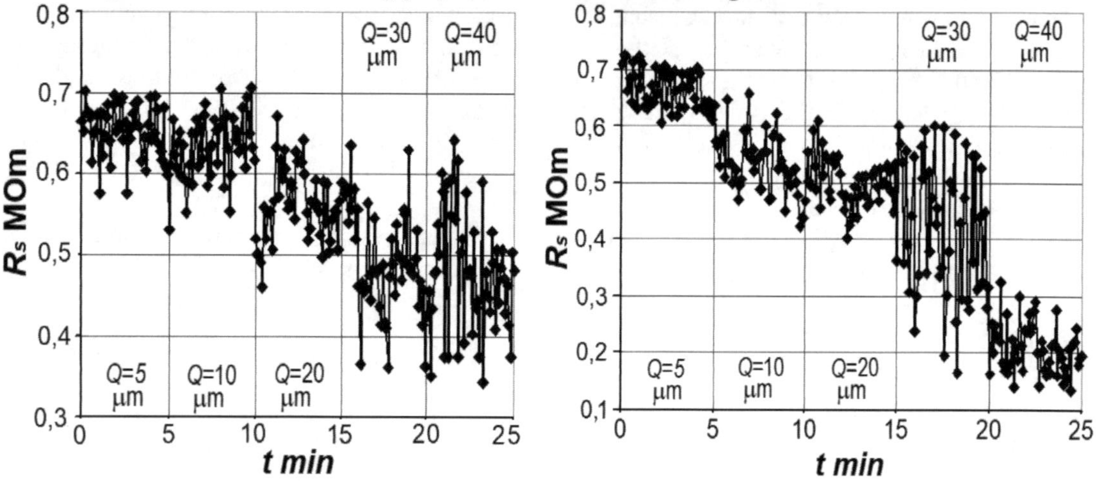

a) b)

a – transmission oil TAD-17; b – plastic lubricant "Litol-24"

Figure 2. Dependence of the R_s parameter on the ovality of a locally loaded ring

In order to confirm the possibility of determining the type of macrodeviations of a locally loaded bearing ring (eccentricity, ovality, faceting), a harmonic analysis of the function of the average resistance of the rolling support from the angle of rotation of the shaft – the function $R_s(\varphi)$ – was carried out. The results of harmonic analysis are shown in Figure 3.

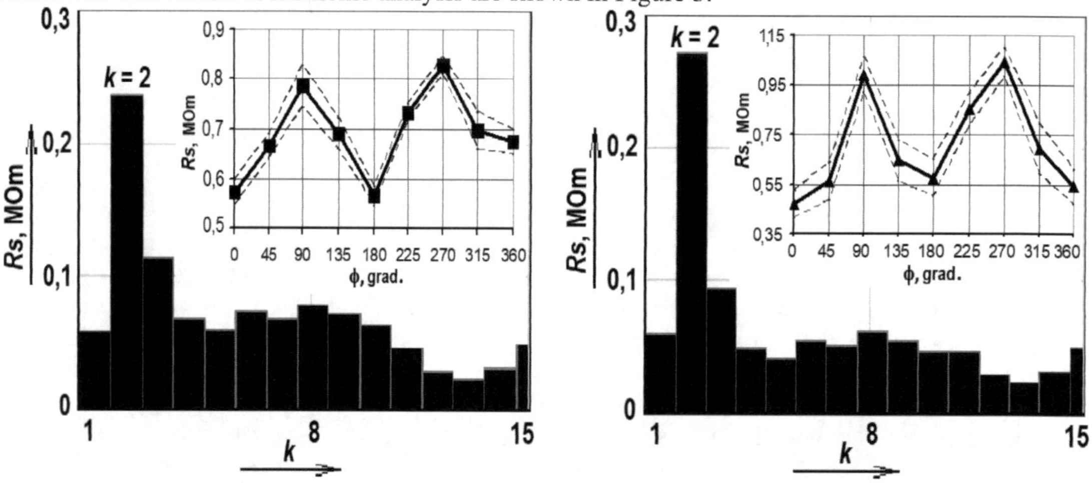

a) b)

a – transmission oil TAD-17; b – plastic lubricant "Litol-24"

Figure 3. Harmonic function analysis $R_s(\varphi)$ for locally loaded ring

Figure 3, a show the results of the harmonic analysis of the $R_s(\varphi)$ function for a locally loaded bearing ring lubricated with TAD-17 gear oil. Figure 3, b shows the results of harmonic analysis of the function $R_s(\varphi)$ for a locally loaded bearing ring lubricated with Litol-24 grease. The spectra of the function $R_s(\varphi)$ were obtained when simulating the ovality of the outer ring of the bearing, equal to Q = 10 µm. For two types of lubricants, the dependences have a pronounced second harmonic of the function $R_s(\varphi)$, corresponding to the ovality of the ring.

Figure 4 shows the results of evaluating the macrodeviations of rolling bearing parts with circulating loading by the diagnostic parameter K_n. During the experiment, a 1000900 type bearing was installed in the rolling support with the number of balls – 7 pcs. For the outer ring of the bearing, ovality was simulated in the range of values from 0 to 40 μm with a step of 10 μm. The inner ring of the bearing was rotated at a frequency of 12.14 Hz (4500 min^{-1}), a circulating load (100 N) created an unbalance device rotating with the shaft.

Figure 4, a show the results of a harmonic analysis of the time function of electrical resistance $R(t)$ of a circulatingly loaded bearing ring lubricated with TAD-17 transmission oil. Figure 4, b shows similar results for a bearing lubricated with Litol-24 grease. Both spectra were obtained with the ovality of the outer ring of the bearing Q2 = 10 μm.

The spectra of Figures 4, a and b show the presence of harmonic components in the function $R(t)$ at information frequencies calculated by the block of formulas (9) and corresponding to the ovality of the outer ring of the bearing.

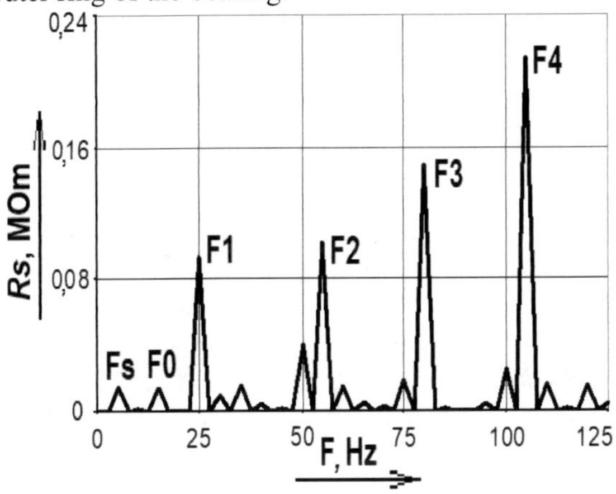

a)

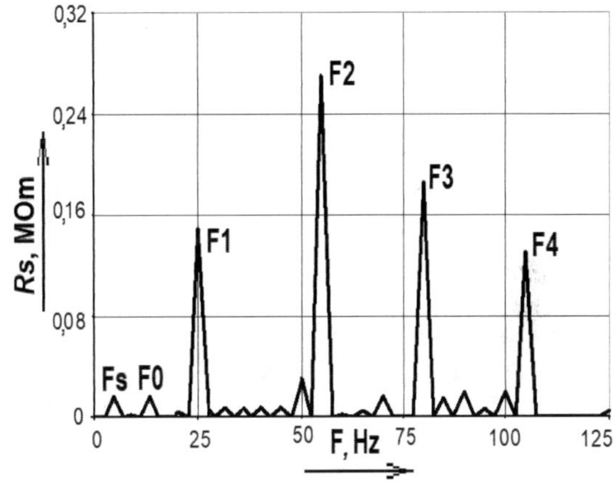

b)

Information frequency values:

Fs=4,5 Hz F0=12,14 Hz
F1=F0·M·L=12,14·2·1=24,3 Hz
F2=Fs·z+F0·M·L=4,5·7+12,14·2·1=55,8 Hz
F3=Fs·z+F0·M·L=4,5·7+12,14·2·2=80,1 Hz
F4=Fs·z+F0·M·L=4,5·7+12,14·2·3=104,4 Hz

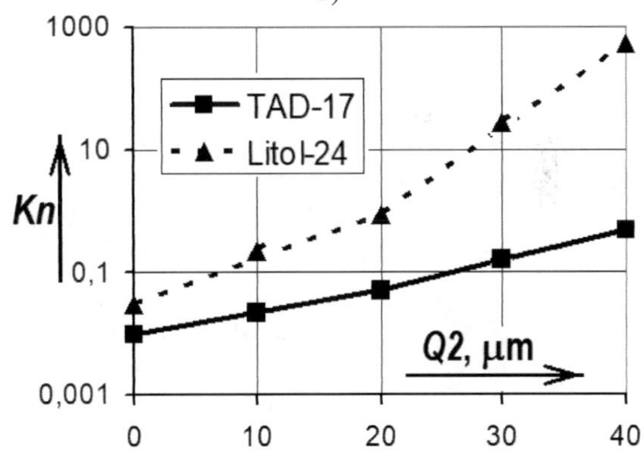

c)

a – transmission oil TAD-17; b – plastic lubricant "Litol-24";
c – parameter K_n dependence from ovality values $Q2$ bearing rings

Figure 4. Analysis of the resistance function for a circulatingly loaded ring

Figure 4, c shows the graphs of the dependence of the diagnostic parameter K_n on the ovality of the outer ring of the bearing in the range of values from 0 to 40 μm with a step of 10 μm. The outer ring of the bearing was subjected to a circulating load (100 N) from the unbalance device rotating with the shaft. The inner ring of the bearing rotated at 12.14 Hz (4500 min^{-1}). The dependences were obtained for two types of lubricants: TAD-17 transmission oil and Litol-24 grease.

The experimental dependences shown in Figure 4, c, confirm the sensitivity of the diagnostic parameter K_n to macrodeviations of the bearing ring under circulating loading. The graph also shows that the parameter K_n has a relatively high sensitivity to the values of macrodeviations – at least two orders of magnitude of the dynamic range of values, the nature of which depends on the value of macrodeviations, and the degree depends on the type of lubricant. These results confirm the theoretical possibility of assessing the macrodeviations of rolling bearing parts with circulating loading by the diagnostic parameter K_n.

The experimentally established monotonic character of the investigated dependences shows that deviations of the shape on the raceway of a circulatingly loaded ring lead to an unambiguous deterioration in the state of the lubricant in the bearing. Spectral analysis of a number of realizations of the function $R(t)$ under different operating modes of the bearing and modeling of shape deviations of various types and values showed that the deviations of the raceway of a circulatingly loaded ring, along with a change in the integral characteristics of the function $R(t)$, cause the formation of information harmonic components. In this case, each type of form deviation corresponds to a set of information components characteristic only for it. Changing the value of the deviation of the form only redistributes the weight of the various components, and their information frequencies remain unchanged.

5. Conclusions on work

The results obtained confirm the reliability of the theoretical studies and the efficiency of the rolling bearing monitoring method by the parameters of the electrical resistance of the bearing assembled in this bearing.

Acknowledgments

The authors consider it their pleasant duty to express their gratitude to the faculty and technical specialists of the Department of Instrumentation, Metrology and Certification of Oryol State University named after I.S. Turgenev for useful participation, great technical assistance provided during the experiments, moral support and valuable advice.

References

[1] Podmasteryev K V 2001 *Electroparametric methods of complex diagnostics of rolling bearings* / K.V. Podmasteryev. M .: Mashinostroenie-1. 376 p.
[2] Markov V V 2004 *Control of rolling bearings by parameters of electrical resistance* / V V Markov, V V Mishin // Control. Diagnostics. N. 9. P. 35.
[3] Podmasteryev K V 2005 *Investigation of the influence of macrodeviations of raceways of rings on the state of lubrication in a bearing by the electroresistive method* / K V Podmasteryev, V V Mishin, V V Markov // Friction and Wear. T. 26. N. 5. P. 546-553.
[4] Demkin N B 2005 *Theory of contact of real surfaces and tribology* / N B Demkin // Friction and Wear. T. 16. N. 6. P. 1003-1024.
[5] Kodnir D S 1976 *Contact hydrodynamics of lubrication of machine parts* / D S Kodnir // M.: Mechanical Engineering. 304 p.
[6] Myshkin N K 2002 *Tribology. Principles and applications* / N K Myshkin, M I Petrokovets. - Gomel: IMMS NASB. 310 p.
[7] Kovalev M P 1980 *Calculation of high-precision ball bearings* / M P Kovalev, N Z Narodetsky // 2 ed. M.: Mechanical engineering. 373 p.
[8] Galakhov M A 1988 *Calculation of bearing assemblies* / M A Galakhov, A N Burmistrov. M.: Mashinostroenie. 272 p.

IOP Conf. Series: Materials Science and Engineering 1211 (2022) 012022 doi:10.1088/1757-899X/1211/1/012022

Color sensor application in high-temperature spectral ratio pyrometer

A A Valke, D G Lobov and A G Shkaev

Omsk state technical university (OmSTU) Mira pr., 11, Omsk, 644050, Russian Federation.

Abstract. *Contactless thermal control tools play an important role in solving the high-temperature technological processes improving energy efficiency problems. In order to create such controls, the authors analyzed the developing possibility of spectral ratio high-temperature pyrometer using a multispectral radiation receiver (color sensor) TCS34725. In the paper this receiver application coefficients are determined, signals ratio graphs in different spectral intervals on temperature are given for two applications: without additional filtration of the control object radiation infrared component and using an opaque in the infrared spectrum part external filter.*

1. Introduction

Currently, there are many technological processes with operating temperature ranges from 1000 to 3000 ° C. The such processes energy efficiency improving is an urgent task today, since this leads to significant savings in energy resources while reducing of the harmful impact relevant productions on the environment.

Such technological processes are found in the metallurgical industry, in the non-combustible thermal insulation production [1] and in carbon black production [2]. Contactless temperature control methods using pyrometers play an important role in solving the problems of the high-temperature technological processes energy efficiency increasing.

2. Problem statement

Two main pyrometers' types are used to measure high temperatures: brightness (or radiation) and colour (or spectral ratio pyrometers). Radiation pyrometers are based on the object radiation energy brightness dependence in a limited wavelength range on its temperature. The spectral ratio pyrometers operation principle is to determine the temperature relative to the signals from several receivers operating in different spectral ranges.

The spectral ratio pyrometers application is preferable in some cases, since the readings of such devices are least affected by viewing windows contamination and by control object viewing channels partial dusting [3-6].

Currently, there are a large number of spectral ratio pyrometers used in production. Such devices include, for example, "Thermokont-TC5P", "S-3000.1 "Steel"", "Raytek MR1S" and others. The main spectral ratio pyrometers disadvantage is the high (hundreds of thousands rubles) cost, a significant part of which is the multispectral radiation receiver (RR) cost.

The article considers the low-cost spectral ratio pyrometer creating possibility, which uses an inexpensive TCS34725 type multispectral radiation receiver (color sensor) the manufactured by AMS-TAOS USA Inc. (USA) [7-9].

3. Theory

It is known that physical bodies with temperatures above absolute zero emit radiant energy in the electromagnetic waves form. At the same time, the radiation power from a surface unit and the radiation spectrum depend on the body temperature. The electromagnetic radiation spectrum for a perfectly black body (PBB) is described by Planck's law:

$$W_\lambda(\lambda,T)=C_1\lambda^{-5}\left(\exp(\frac{C_2}{\lambda T})\right)^{-1}, \tag{1}$$

where W_λ is the flow spectral density, C_1 and C_2 are constants, λ is the wavelength, T is the body temperature.

$$c_1=2\cdot\pi\cdot h\cdot c^2 ; \tag{2}$$

$$c_2=\frac{h\cdot c}{k}, \tag{3}$$

where c is the speed of light in vacuum, h is Planck's constant, k is Boltzmann's constant.

The graphs in Figure 1 show the change in the PBB radiative flux relative spectral density W_λ for different temperatures.

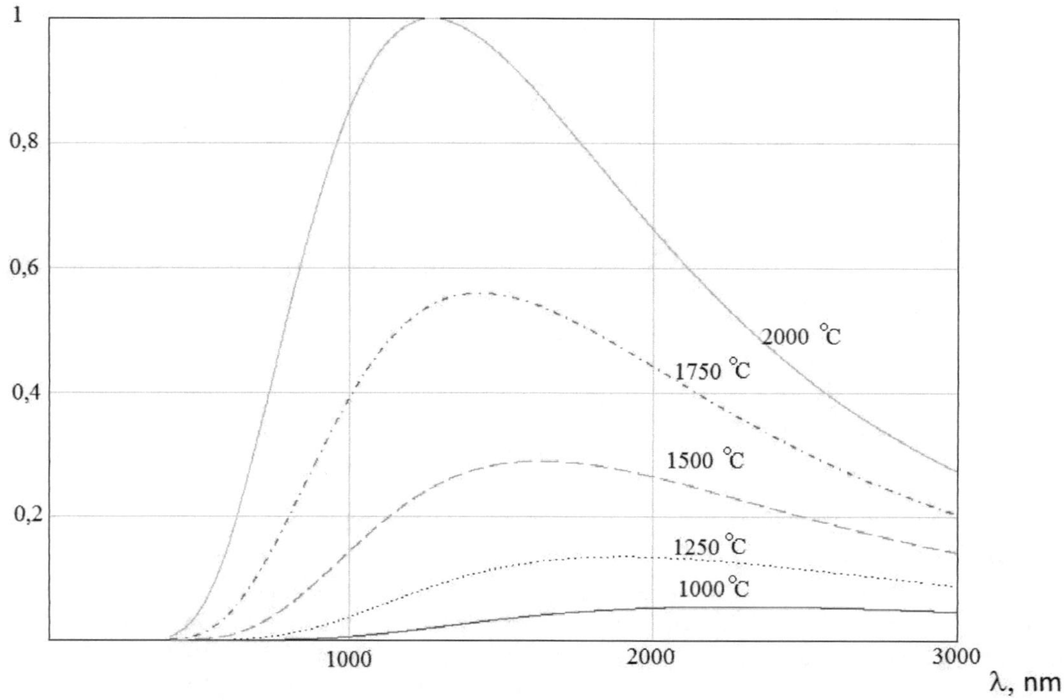

Figure 1. Change of PBB radiative flux relative spectral density W_λ for different temperatures.

Figure 1 shows that the Planck function maximum in the temperature range from 1200 to 2000 ° C is in the wavelength range of 1-2 microns. At the same time, when the temperature rises, the maximum shifts to the shorter-wavelength part of the spectrum. Accordingly, the radiation receiver should work in this spectrum part. The TCS34725 color sensor normalized spectral characteristic is shown in Figure 2. [10]

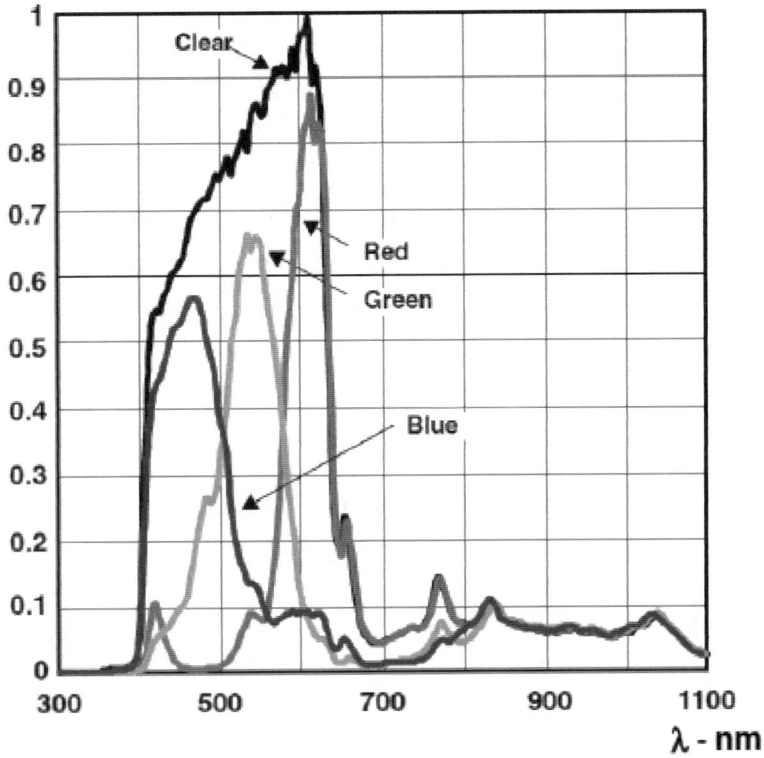

Figure 2. TCS34725 color sensor normalized spectral characteristic.

Figure 2 shows that the R-channel maximum spectral characteristic is at 615 nm, the G-channel maximum spectral characteristic is at 525 nm, and the B-channel maximum spectral characteristic is at 465 nm. During the research, the TCS34725 color sensor use possibility as a radiation receiver for measuring temperatures in the range from 1200 to 2100 ° C was considered. This temperature range is typical for the carbon black production technological process.

4. Experimental results

During the research, the radiation receivers use coefficients for each of the three TCS34725 color sensor channels were calculated first, for 1200 ° C and 2100 ° C temperatures.

The use factor is determined by the dependence [4]:

$$K = \int_0^\infty \Phi(\lambda) \cdot S(\lambda) d\lambda \,/\, \int_0^\infty \Phi(\lambda) d(\lambda) \qquad (4)$$

where $\Phi(\lambda)$ is a function describing the distribution of the flow from the wavelength falling on the RR at a given controlled surface temperature; $S(\lambda)$ is the RR sensitivity spectral characteristic. Thus, this coefficient shows what radiative flux proportion falling on the RR corresponds to its spectral sensitivity, and, therefore, is converted into an output signal.

In this case, the function Φ (λ) for the given temperatures was calculated in accordance with Planck's law, and the spectral characteristics functions $S(\lambda)$ of the radiation receivers R-, G-, and B-channels were determined by the normalized spectral characteristics spline interpolation shown in Figure 2.

The use factor calculated values for different channels for temperatures of 1200 °C and 2100 ° C are shown in Table 1.

In contradistinction from a brightness pyrometer, a spectral ratio pyrometer determines the object temperature with respect to signals from several receivers operating in different spectral ranges. Therefore, the signals ratio from channels R, G and B was calculated in the measured temperatures range from 1200 to 2100 °C.

Table 1. Use coefficients.

Channel	Temperature	
	1200 °C	2100 °C
R	0.073	0.1
G	0.067	0.085
B	0.066	0.07

Figures 3 and 4 show the signals ratio dependence of channels G and R, and channels B and R on temperature without using an external additional filter.

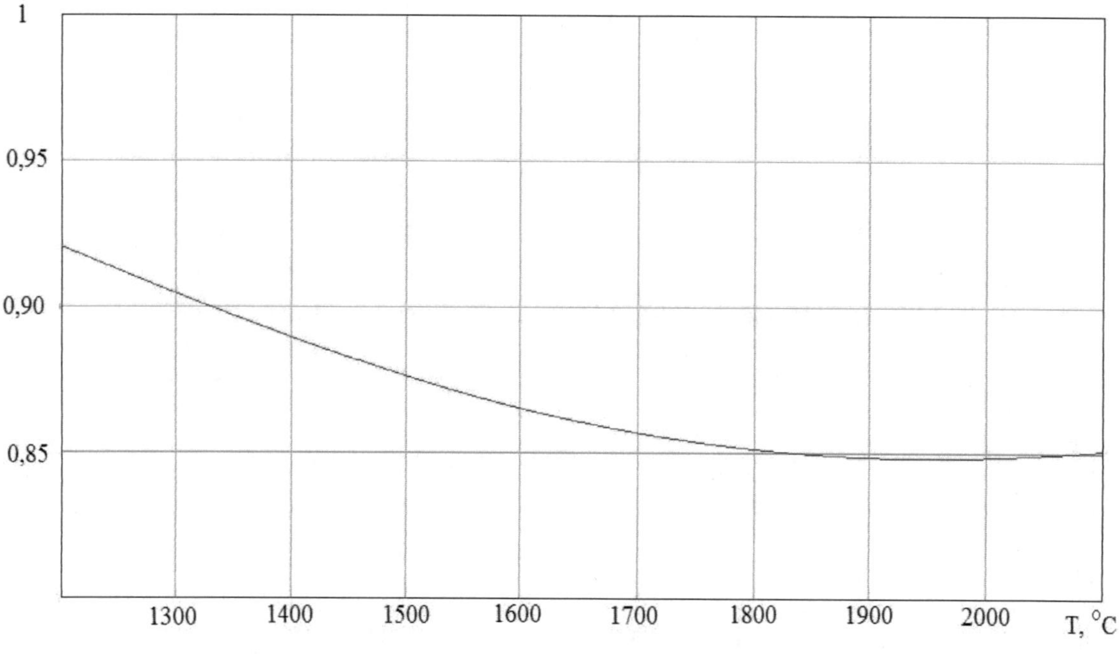

Figure 3. The signals ratio dependence of channels G and R on temperature without using a filter.

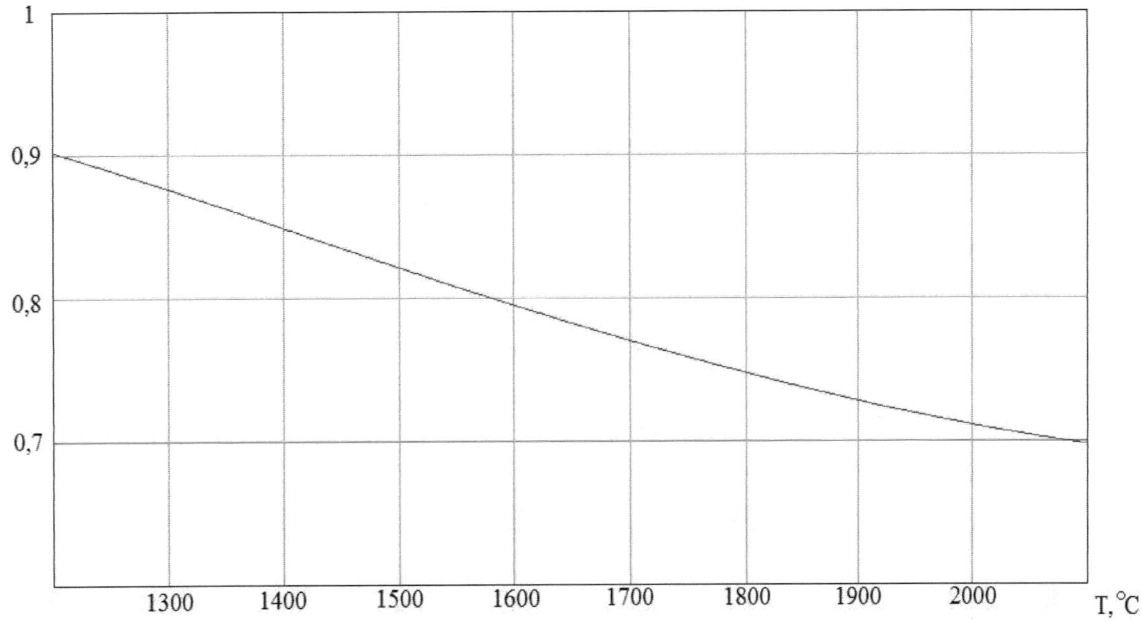

Figure 4. The signals ratio dependence of channels B and R on temperature without using a filter.

Figures 5 and 6 show the signals ratio dependence of channels G and R, and channels B and R on temperature using an external filter that cuts off the infrared spectrum part at a wavelength of 700 nm.

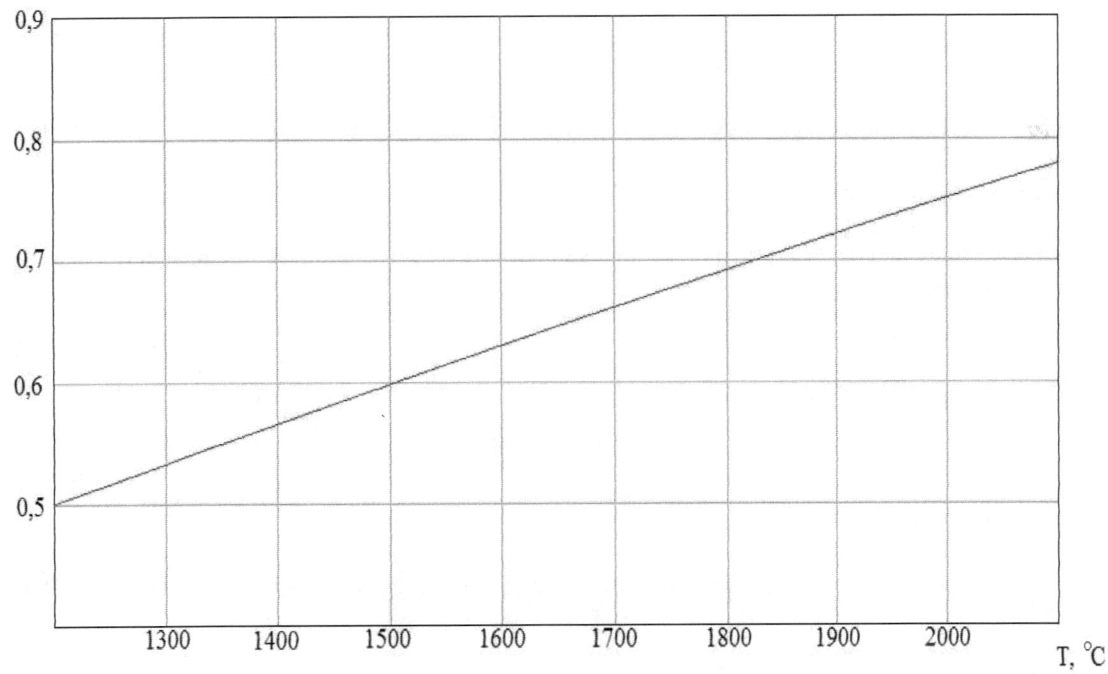

Figure 5. The signals ratio dependence of channels G and R on the temperature with the filter.

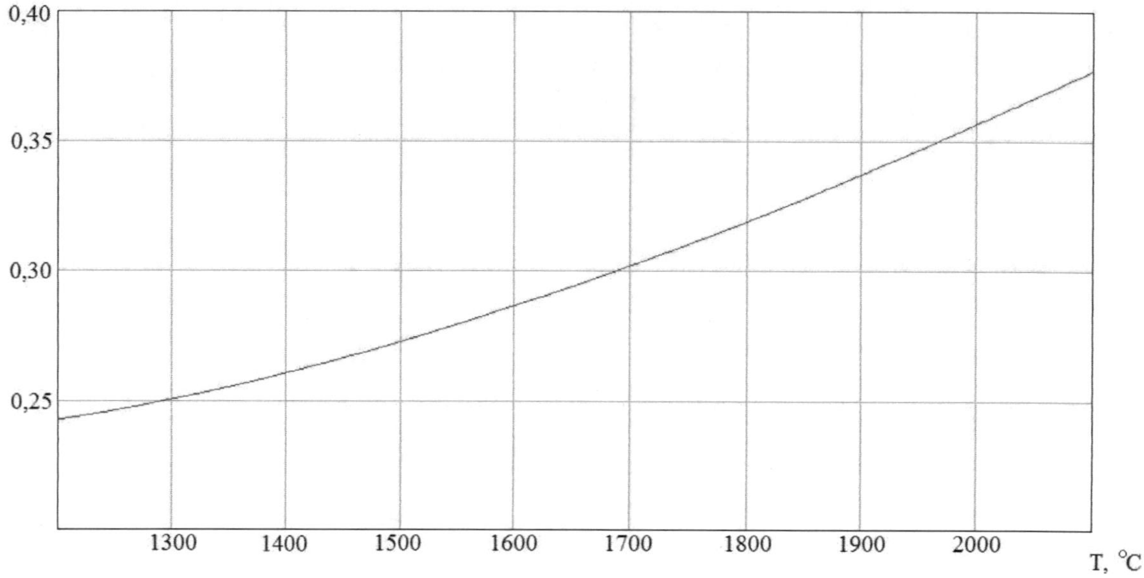

Figure 6. The signals ratio dependence of channels B and R on the temperature with the filter.

Based on the presented results, inference should be drawn that the TCS34725 color sensor can be used as a radiation receiver for a spectral ratio pyrometer with a temperature range of 1200 - 2100 ° C. In this case, the best option is to make a spectral ratio pyrometer using channels G and R with an external filter that cuts off the infrared spectrum part.

5. Conclusion

This solution, in accordance with Table 1, provides the highest radiation receivers use coefficients, and at the same time, the signals ratio dependence of the channels G and R on temperature is monotonically increasing and almost close to linear (Figure 5). These two factors make it possible to simplify the spectral ratio pyrometer electronic scheme implementation.

References

[1] Valke A A, Lobov D G, Shkaev A G, Shkaev A A and Ponomarev D B 2021 Visual level measurement system application in the technological process automated control system of the recuperative furnace bath, *J. of Phys:* **1791**

[2] Sun D L, Wang F, Hong R Y and Xie C R 2016 Preparation of carbon black via arc discharge plasma enhanced by thermal pyrolysis *Diam. Relat. Mater.* **61** pp 21–31

[3] Zhang Y, Lang X, Hu Z and Shu S 2017 Development of a CCD-based pyrometer for surface temperature measurement of casting billets *Measurement Science and Technology* **28**

[4] Wang W, Lin J, Zhong W, Zhang B, Xu C and Ding H 2018 Analysis of infrared temperature measurement for flu gas shielding metal surface using source multi-flux method *Thermal science* 1A **22** pp. 313-321.

[5] Srsen D Temperature measurement and thermal control of die casting processes using CCD cameras 2012 *Electronic Theses and Dissertations* 93.

[6] Tairan F, Huan Z, Jun Z, Zhe W, Maohua Z and Congling S. 2010 Improvements to the three-color optical ccd-based pyrometer system *Applied Optics* pp. 5997–6005.

[7] Mendeleev V, Petrov V, Yashin A, Vangonen A and Taganov O 2020 Application of a pyrometer and standard sample to determine the surface temperature of materials under study *Measurement Techniques* **12** Vol. 62 pp. 1029-34.

[8] Sun X, Dai J, Cong D and Coppa P 2002 Development of a special multi-wavelength pyrometer for temperature distribution measurements in rocket engines *International Journal of Thermophysics* **5** Vol. 23 pp. 1293-1301.

[9] Kropachev D, Grishin A and Maslo A 2012 Real-time methods of measuring the temperature of metallic melts at machine plants. *Metallurgist* **5-6** Vol. 56 pp. 472-474.

[10] TCS3472 color light-to-digital converter with IR filter 2012 *Reference manual* 26

The Diagnostics of the rotor misalignment of a permanent magnet motor

A A Afanasev[1], V S Genin[1], L N Vasileva[1] and V G Grigorev[2]

[1]Department of Automation and Management in Engineering Systems, Chuvash State University named after I.N. Ulyanov, 15, Moskovsky prospect, Cheboksary, Russia
[2]Department of Industrial Electronics, Chuvash State University named after I.N. Ulyanov, 15, Moskovsky prospect, Cheboksary, Russia

Abstract. A mathematical model of the magnetic field in the working gap of a brushless motor is considered in a case of rotor misalignment arising during manufacture, for example, due to defects in end shields, or in operation due to bearing wear. a gap in a uniform (circular ring). The stator gearing is taken into account on average using the Carter coefficient, the magnetic field in the inhomogeneous air gap, created by the rotor magnets and the stator winding current, is assumed to be plane-parallel, having a two-dimensional character. It was found that the rotor misalignment associated with the rotational movement of the eccentricity causes nonsinusoidality of the idle EMF and pulsation of the electromagnetic moment with a frequency 3p times higher than the rotor speed. When the eccentricity is stationary, a variable EMF is induced along the rotor shaft, causing an alternating current in the circuit: shaft-bearings-bearing shields-stator housing. To clarify the nature of the defect in order to identify the actual misalignment of the rotor, it is recommended to control currents and voltages using specialized software and hardware complexes for spectrum analysis.

Electrical machines are widely used in various industries. They are the largest consumers, using up to 80% of all electricity generated in the country. In recent decades, thanks to the development of semiconductor power and microprocessor technology, permanent magnet motors have been successfully used. The main advantages of drives based on such electric motors:

- high short-term overload torque, reaching 10 times the nominal value;
- high energy indicators ($\cos\varphi = 1$, $\eta = 0.9 \div 0.98$);
- significantly smaller weight and dimensions.

In particular, permanent magnet ac converter-fed motors are widely used in drives of automation and robotics systems, in vehicles.

Often, as a result of untimely identification of developing equipment defects, the likelihood of severe consequences in accidents increases, and the volume of repair work increases. In many industries, sudden engine failure can lead to irreparable consequences. In this regard, an important condition for the effective operation of electrical equipment is diagnostics through constant monitoring and assessment of the condition. This is especially true when using equipment that has reached its standard service life. Diagnostic monitoring of equipment also creates the prerequisites for the transition from a system of preventive maintenance to maintenance based on technical condition or maintenance, focused on reliability (RCM - Reliability Centered Maintenance), reducing operating costs.

The rotor misalignment of the brushless motor can occur during manufacture, for example, due to defects in end shields, or in operation due to bearing wear. In the first case, the section of the air gap with the minimum size (eccentricity) is stationary, associated with a specific place on the surface of the stator bore, in the second, it will rotate with the rotor. It should be noted that damage to bearing elements, along with turn-to-turn faults in the stator winding, is one of the most common types of damage.

In [1], using the conformal transformation of an uneven air gap in the plane z into a uniform (circular ring) in the plane t (Figure 1), the calculations of the magnetic field in the air gap of the engine were performed. When considering the phenomena in the engine caused by rotor misalignment, the following assumptions were made:
1) the magnetic permeabilities of the stator core and the rotor yoke are equal to infinity;
2) the gearing of the stator is taken into account on average using the Carter coefficient;
3) the magnetic field in the inhomogeneous air gap, created by the rotor magnets and the stator winding current, is plane-parallel, has a two-dimensional character.

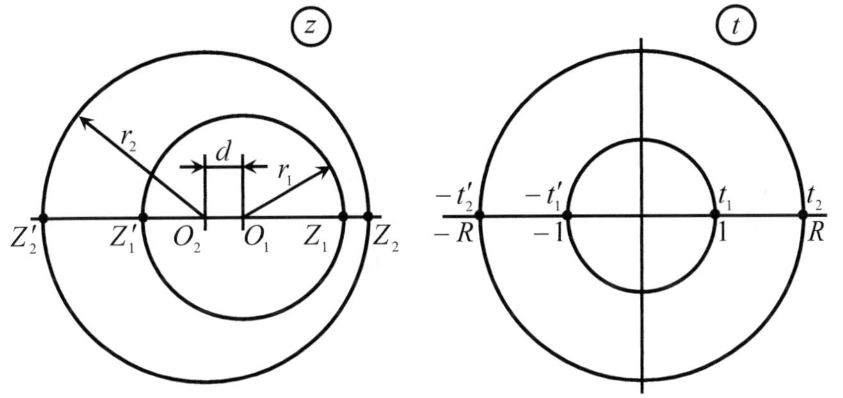

Figure 1. Conformal transformation of an uneven air gap into a uniform one.

Conformal transformation is performed according to the formula [2]

$$t = \frac{(A+1)z - (z_1'A + z_1)}{(A-1)z - (z_1'A - z_1)}, \quad (1)$$

where $A = \sqrt{\dfrac{r_2^2 - (d+r_1)^2}{r_2^2 - (d-r_1)^2}}$; z_1, z_1' – diametrical points of the inner circle; r_1, r_2 -the radiuses of the inner and outer circumferences; it is an offset of the d - centers of the circles.

The field in the annular region of the complex plane t is defined as the solution of the first boundary value problem with known scalar magnetic potentials $t = f_1(\phi)$ and $t = f_2(\phi)$, respectively, on the inner (unit radius) and outer (radius R) circles. Since the scalar magnetic potential under conformal transformations is an invariant quantity [3], the magnetic potential of the rotor magnets is represented by the expression

$$f_2(\phi) = \frac{4}{\pi} H_{cB} h_{\text{M}} \sum_{k=1}^{\infty} \frac{\cos(2k-1)p\alpha}{2k-1} \sin(2k-1)p\phi, \quad (2)$$

where H_{cB}, h_{M} – the coercive force of induction and height of magnets; p – number of pairs of motor poles; α – the half of the angular distance between adjacent oppositely polarized magnets in radians.

The calculations were carried out in relation to the 6DVM300 A35 ac converter-fed motor (see Figure 2) [4], the serial production which was mastered by ChEAZ JSC (Cheboksary): $D_i = 126\,\text{mm}$, $l = 150\,\text{mm}$, $h_м = 7{,}1\,\text{mm}$, $H_{cB} = 800\,\text{kA/m}$, $M = 70\,\text{Nm}$.

Figure 2. Ac converter-fed motor 6DVM300.

The dependence of the scalar potential on the inner boundary of the circular ring of the plane t, the magnetic potential and radial induction inside the circular ring, etc. are obtained.

With the help of the conformal transformation (1), a connection is established between the magnetic inductions in the complex planes z and t [5]

$$B(z) = B(t)\left|\frac{dt}{dz}\right|. \tag{3}$$

where

$$\left|\frac{dt}{dz}\right| = f(z) = \frac{(A+1)\left[(A-1)z-(z_1'A-z_1)\right]-(A-1)\left[(A+1)z-(z_1'A+z_1)\right]}{\left[(A-1)z-(z_1'A-z_1)\right]^2} \tag{4}$$

The radial components of the magnetic inductions at the boundaries of the air gap, obtained according to (3) and (4), are shown in Figure 3. Due to the displacement of the inner circle in the direction of the radius with a zero angular coordinate (Figure 1), at values close to the maximum values $\phi = 0\ (2\pi)$ are obtained induction maxima. Oscillations of induction on the graph are explained by the use of a limited number of terms of an infinite series.

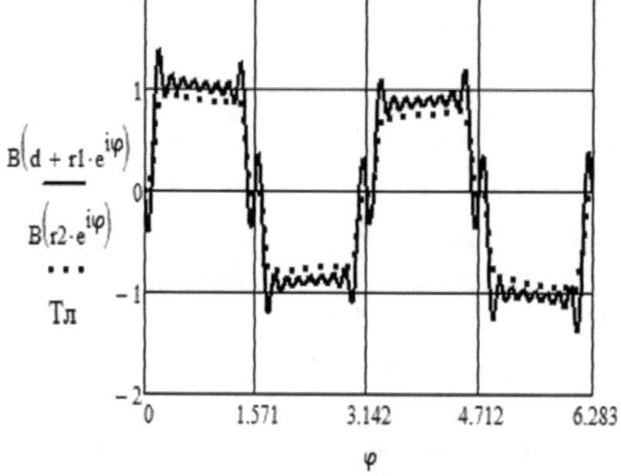

Figure 3. Values of magnetic inductions at the boundaries of the air gap in the plane.

Then, the amplitude values $L_{Af\nu}$ of the inductances of mutual inductance of the rotor magnets with the phase A of the stator winding are determined, taking into account that the minimum air gap rotates with the rotor

$$L_{Af\nu} = \frac{4}{\pi^2} \frac{wk_w(\nu)\tau l}{H_{cB} h_M} \int_0^{2\pi} B(r_2 e^{j\phi}) \sin\phi \, d\phi , \tag{5}$$

with which you can find the flux linkage $\psi_A(\phi)$ and the idle EMF $e_{A0}(\phi)$ of this phase

$$\psi_A(\phi) = H_{cB} h_M \sum_{\nu=1}^{\infty} L_{Af\nu} \sin \nu\phi , \tag{6}$$

$$e_{A0}(\phi) = -\frac{d\psi_A(\phi)}{dt} = \frac{1}{2} \omega H_{cB} h_M \sum_{\nu=1}^{\infty} L_{Af\nu} \cos \nu\phi , \tag{7}$$

where $\omega = \frac{2d\phi}{dt}$ – the electric angular is frequency of the considered four-pole motor.

Figure 4 shows the idle EMF caused by the rotor magnets, calculated according to (6), (7) for a 6DVM300 A35 ac converter-fed motor.

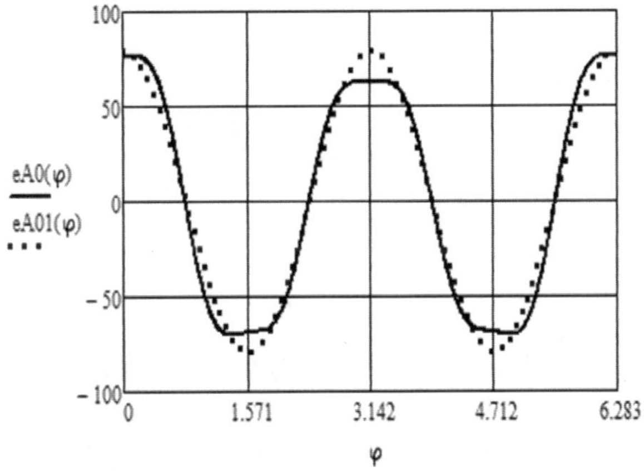

Figure 4. No-load EMF of phase A (resulting e_{A0} and fundamental harmonic e_{A01}) due to the rotor magnets.

We suppose that a permanent magnet motor has almost sinusoidal current and voltage in the stator winding, the phase angles of which can be linked by hardware and software to the angular position of the rotor.

The electromagnetic torque of the motor is determined using equality (7) and assuming that the phase currents are in antiphase with the no-load EMF

$$M(\phi) = \frac{1}{2} \frac{p}{\omega} \begin{bmatrix} e_{A0}(\phi) i_A(\phi) + e_{A0}(\phi - 2\pi/3) i_A(\phi - 2\pi/3) + \\ + e_{A0}(\phi - 4\pi/3) i_A(\phi - 4\pi/3) \end{bmatrix}, \tag{8}$$

where $i_A(\phi) = I_m \cos p\phi$.

The graphic of the change in the electromagnetic moment, built according to (8), is shown in Figure 5. We see that with a rotor misalignment, for example, 1 mm, an alternating component of the electromagnetic moment appeared, which has a frequency six times higher than the frequency of the stator current, and the ripple depth of which at rated current is 7.1% of rated torque.

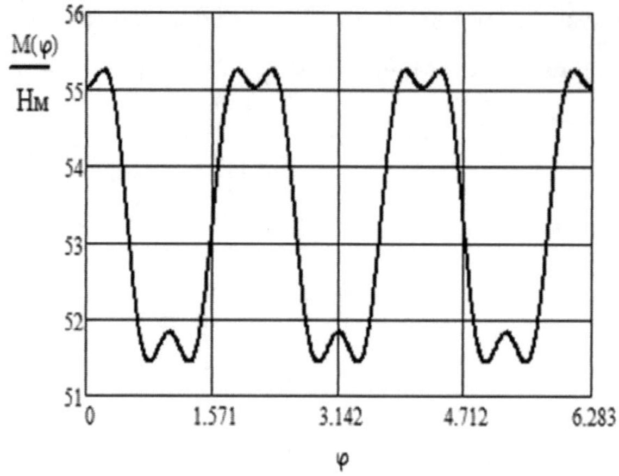

Figure 5. Electromagnetic moment at rated stator current as a function of the angular position of the rotor.

If the eccentricity of the rotor is stationary relative to the stator surface, the symmetric magnetic field in the circular ring of the plane will rotate, and the function $f(z)=\frac{dt}{dz}$, connecting the values t and z of induction in planes and, remains stationary. As a result, the magnetic field in the poles (magnets) and the rotor yoke will begin to pulsate, and the total alternating magnetic flux, closed around the rotor shaft, will be nonzero. In this case, as is known [6], an EMF is induced along the shaft and an alternating current flows, closing through the bearings, end shields and the stator housing. As a result, bearing heating may occur, accompanied by electrochemical erosion.

From the foregoing it follows that the excess of the permissible values of the rotor misalignment of the brushless motor can be detected by monitoring the voltage at the motor terminals, the consumed current and the temperature. However, the asymmetry and non-sinusoidality of phase currents and voltages, accompanied by an increase in temperature, can be caused by other reasons, for example, turn-to-turn faults in the stator windings, magnetic asymmetry of the rotor caused by inhomogeneous magnetization of magnets, and other reasons.

Thus, monitoring of temperature, currents and voltages may indicate a malfunction. To clarify the nature of the defect, it is possible to recommend vibrodiagnostics of the state of the engine elements [7], which, however, is a rather expensive and time-consuming measure. Recently, methods for diagnosing electrical machines have been developed based on the spectral analysis of the current signal [8-10]. The use of specialized software and hardware complexes [11] makes it possible to determine the state of various engine elements with a high degree of reliability.

Conclusions

1. Misalignment of the rotor of a brushless motor can occur during manufacture, for example, due to defects in end shields, or in an operation due to wear of bearings, which happens very often.
2. The misalignment of the rotor, associated with the rotational movement of the eccentricity, causes nonsinusoidality of the idle EMF and pulsation of the electromagnetic moment with a frequency of $3p$ times the rotor speed. When the eccentricity is stationary, a variable EMF is induced along the rotor shaft, causing an alternating current in the circuit: shaft-bearings-bearing shields-stator housing.
3. To clarify the nature of the defect in order to identify the actual misalignment of the rotor, it is possible to recommend monitoring currents and voltages using specialized software and hardware complexes for spectrum analysis.

References

[1] Afanasev A A 2020 *Mathematical Modeling of Electromechanical Systems (*Cheboksary: Chuvash State Univesity) p 274

[2] Koppenfels V and Shtalman F 1963 *Practice of Conformal Mappings* (Moscow: Izd-vo inostrannoi literatury) p 407

[3] Polyanin A.D. 2001 *A Handbook of Linear Equations in Mathematical Physics* (Moscow: Fizmatlit) p 576

[4] *Rules for the Citing of Sources* Available from: https://www.cheaz.ru/assets/images/production/10-ensber/1-electrodvig/2-6-dvm/6dvm.pdf

[5] Ivanov-Smolenskii AV and Abramkin YuV 1985 *Application of Conformal Transformation in Electrical Machine Calculations for Electrical Machines. Analytical Methods* (Moscow: Tipografiya MEI) p 85

[6] Alekseeva MM 1967 *Machine Generators of Increased Frequency* (L.: Energiya) p 344

[7] *Condition Monitoring and Diagnostics of Machines. Vibration Monitoring of the State of Machines Part 9. Methods for Diagnosing Electric Motors* (ISO/DIS 13373-9, YuT)

[8] Thomson W T and Rankin D1987 *Case Histories of Rotor Winding Fault Diagnosis in Induction Motors* 211 Int Conf Proc on Condition Monitoring, University College Swansea

[9] Thomson W T and Gilmore R J 2003 *Motor Current Signature Analysis to Detect Faults in Induction Motor Drives Fundamentals, Data Interpretation, and Industrial Case Histories* Discipline Technical Authority (Electrical Engineering) AMECU pstream Oil&Gas Nigg, Aberdeen, Scotland. Available from: http://docplayer.net/21734312-Motor-current-signature-analysis-to-detect-faults-in-induction-motor-drives-fundamentals-data-interpretation-and-industrial-case-histories.html

[10] ISO20958:2013 (en) *Condition monitoring and diagnostics of machine systems – Electrical signature analysis of three-phase induction motors*

[11] Petukhov V S *Diagnostics of the Condition of Electric Motors Based on the Spectral Analysis of the Consumed Current* Available from: http://mcsa.electrik.org/el1.html

AUTHOR INDEX